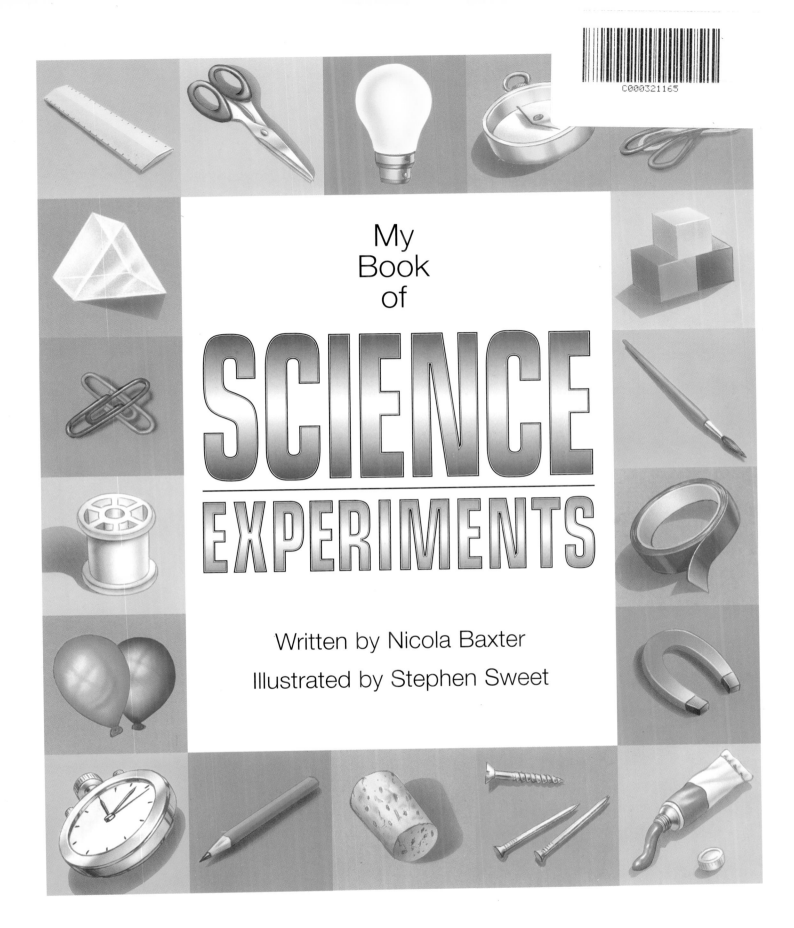

My
Book
of

SCIENCE
EXPERIMENTS

Written by Nicola Baxter

Illustrated by Stephen Sweet

ARMADILLO

This paperback edition printed in 2008

© 2002 Bookmart Limited

ISBN: 978-1-84322-348-1

7 9 10 8 6

Published by Armadillo Books
an imprint of Bookmart Limited,
Registered Number 2372865
Trading as Bookmart Limited
Blaby Road, Wigston
Leicestershire
LE18 4SE

Produced for Bookmart Limited by Nicola Baxter
PO Box 215, Framingham Earl, Norwich NR14 7UR

Designer: Amanda Hawkes

Photographs: Corel Professional Photos
Cover illustrations and artwork on pages
1,2,3,5,6,7,35,40,41,55, 82 and 103 by Duncan Gutteridge

Originally published in hardback
by Bookmart Ltd in 2000

Printed in Thailand

BEFORE YOU EXPERIMENT...

Science experiments are fun and can teach you a lot about the world around you, but good scientists always put safety first.

Remembering these points will help to make sure that your experiments are enjoyable *and* safe.

1 Always ask an adult before you carry out **any** of the experiments in this book. Show him or her the experiment and explain how and where you are going to do it. If the experiment says that an adult needs to help you, make sure that he or she stays for the whole experiment. Never try to use sharp tools or heat substances by yourself.

2 Look at the list of things you will need before you begin. Make sure you have everything ready before you start.

3 There are two kinds of measurements in this book – metric ones and imperial measurements in brackets afterwards. Decide which to use and stick to it – don't use a combination of the two in the same experiment.

4 Always have a notebook and pencil by your side. You may need to write down your results. It is useful to get into the habit of writing down what you think will happen before you do an experiment. Then record your results afterwards and see if you were right. If not, can you work out why?

5 Be very careful when using household chemicals, such as soap or washing-up liquid in your experiments. Take care, too, when you are using food items. Make sure you have permission to use them and throw them away carefully after use. They are not fit for eating after you have experimented with them! Always wash your hands before and after using any chemicals or foods.

6 Clear up everything you have used at the end of an experiment. Always put things away safely. This is especially important with items such as glasses, bottles, scissors, knitting needles, elastic bands and plastic bags, which can be very dangerous for smaller children and pets. Remember that even a small bowl of water can harm a small child.

Abbreviations used in this book

C	Celsius or centigrade
cm	centimetre
F	Fahrenheit
Hz	Hertz
in	inch
km	kilometre
lb	pound
m	metre
oz	ounce
V	volt

Contents

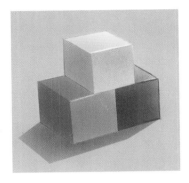

Music and sound

The air around us

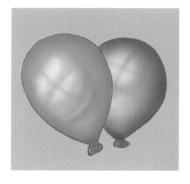

Amazing Materials

Mighty molecules

Everything in the universe is made up of atoms, but atoms do not usually exist singly. They group together in larger structures called molecules. These can be made of atoms of the same kind or of different kinds. A molecule of water, for example, is made of two atoms of hydrogen and one atom of oxygen. These atoms are chemically bonded together.

EXPERIMENT Molecules stick together

You will need
- card
- scissors
- bowl of water
- washing-up liquid

1 Cut a boat shape from card and float it gently on the water.
2 Dip your finger in the washing-up liquid and carefully lower it into the water just behind the boat. What happens?

The molecules on the surface of water cling so closely together that they seem to form a "skin", called surface tension. The detergent decreases the surface tension behind the boat. The stronger surface tension in front pulls the boat forward.

Molecules on the move

You will need
- an empty aquarium or large, clear-plastic container
- strong wire
- hot water
- cold water
- ice cube
- small bottle or jar
- ink

1 Fill the aquarium with cold water and float the ice cube at one end.

2 Fill the small bottle with hot water coloured with a little ink.

3 Twist wire round the top of the bottle and lower it carefully into the aquarium.

Although a liquid may seem to be still, its molecules are actually moving around all the time. Molecules move more quickly when they are warm. They move upwards. Cooler molecules sink. In this way, a current, called a convection current, is formed in the water.

Clever crystals

You will need
- **an adult to help**
- sugar
- water
- saucepan
- pencil
- fine string
- empty yogurt pot

The same molecules can form different shapes. Sugar dissolved in water is liquid. It turns into crystals again as the water evaporates.

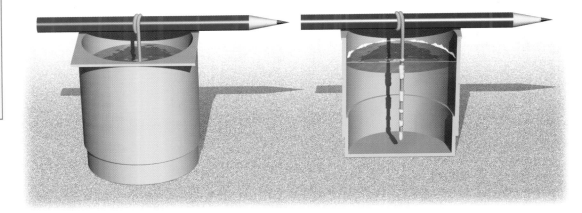

1 Ask an adult to help you heat two parts of water to one of sugar until the sugar has dissolved.

2 When the liquid is cool, pour it into a yogurt pot and arrange a pencil and string as shown. What happens in a few days?

Water ways

One of the remarkable properties of water is that it always tries to reach the lowest point it can. Engineers can use this fact to channel water from one place to another. Like air, water also presses down on anything within it. The greater the depth of water, the greater the pressure. Divers have to be careful of the effects of changes in water pressure on their bodies.

EXPERIMENT The great leveller

You will need
- clear plastic hose or tube
- funnel
- water
- jug

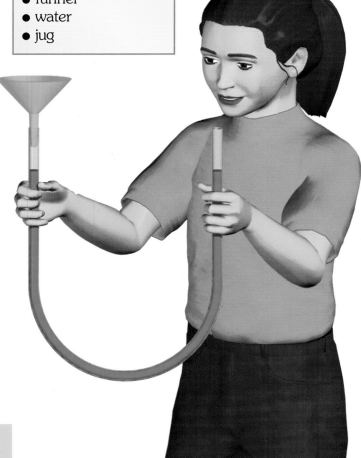

❶ Fit the funnel into one end of the hose and pour some water into it, holding up the other end so that it doesn't flow out.
❷ Now try lifting one end of the hose. What happens to the level of water at both ends? Can you do anything to make the water at one end higher than that at the other?

❸ You can use this property of water to make a fountain. Wear old clothes and go outside first! Squeeze one end of the hose a little to make a nozzle. Then raise the other end as high as you can. It may help to ask a taller friend to do this. The water will spurt out of the lower end. The higher you raise the upper end, the higher the fountain. Once again, it is trying to reach its own level.

Super siphon

You will need

- a friend to help you
- 2 washing-up bowls
- length of hose
- water
- stairs

❶ Half fill one washing-up bowl with water and put it at the top of the stairs. Put the empty bowl at the bottom.

❷ Ask a friend to pinch one end of the hose closed while you fill the other end from the tap. Pinch your end too.

❸ Ask your friend to take her end of the hose up to the top bowl and, still pinching it, put the end under the water.

❹ Put your end in the bowl at the bottom. Tell your friend to stop pinching, then let go yourself. Air pressure pushes the water up the tube so that it can then flow down to the bottom bowl.

Water pressure

You will need

- plastic bottle
- knitting needle or pair of compasses
- water

❶ The further down a diver dives, the greater the water pressure. This pressure can be shown by the force with which water escapes if given a chance. Make three holes in a plastic bottle as shown.

❷ Fill the bottle with water and stand it up straight. The deepest water, escaping from the hole at the bottom, is under greater pressure, so it jets out further.

Fabulous fabric

Cloth (or fabric) is made by weaving, knitting or pressing fibres together. Fibres may be natural, such as cotton, wool, silk and linen, or artificial, made from chemicals. Fabrics have a wide range of properties. This is particularly important when they are used to make clothes – they may need to keep us warm or cool. We may want them to absorb moisture or repel it.

EXPERIMENT Winter warmth

You will need
- samples of cloth
- clean, empty yogurt pots
- water
- elastic bands
- teaspoon

❶ Fill three or four yogurt pots to exactly the same level with water and put them in the freezer for an hour or two.

❷ When the water is completely frozen, take the pots out and wrap each of them in several layers of a different kind of cloth. Use elastic bands to secure them. Use a piece of the same cloth to cover the top of each pot.

❸ After half an hour, unwrap the pots and carefully lift out the lumps of ice inside. Now use a teaspoon to measure how much water is in each pot. The one with least water was the best insulated. Its fabric would help to stop heat escaping from your body on a cold day.

Wet and dry

❶ Test the absorbency of different kinds of cloth by fixing one layer of each kind to the top of a clean, dry yogurt pot with an elastic band. Carefully drip one teaspoon of water on to each pot. Repeat this two or three times.

❷ Remove the cloths from the pots and see what has happened. The pots with very little water inside have very absorbent cloths. They are good at soaking up liquids. The pots that have a lot of water inside or pooled on top have less absorbent cloths.

Waterproofing

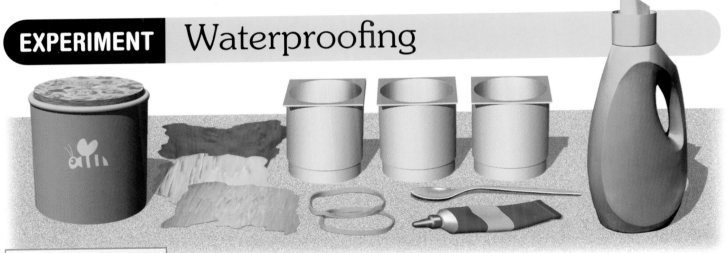

Sometimes we want fabrics to be as waterproof as possible. Absorbent threads can be waxed, oiled or coated with rubber or plastic to achieve this.

❶ Repeat the above experiment but first spread oil, wax or glue on the fabric.

❷ This time, look to see if the water sits on the surface of the fabric without being absorbed. That is what should happen if the fabric is waterproof.

Paper power

Paper is a very versatile material. It can be cut, folded, moulded and recycled. It also provides an excellent surface for printing and coatings of various kinds. Paper is made from plant fibres, mainly from farmed trees. When mixed with water, they form a pulp that can be spread on a mesh. The pulp is pressed and dried, then lifted from the mesh as paper.

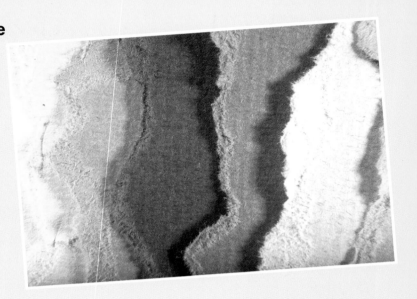

EXPERIMENT | Against the grain

You will need
- newspaper

❶ Try to tear a piece of paper in a straight line from one edge. How easy is it?

❷ Now try tearing from the adjoining edge. On the mesh of a papermaking machine, the fibres in pulp tend to line up lengthwise,

giving the paper a "grain". A tear along the grain will be straighter than a tear across the grain.

EXPERIMENT | Simple sizing

You will need
- blotting paper
- PVA glue
- water
- paintbrush
- felt-tip pens

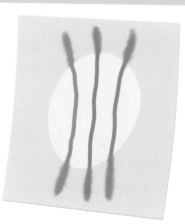

❶ Paper for printing needs a surface that ink will not soak into. "Size" is sprayed on to paper to achieve this. Thin some PVA glue with water and paint a patch of it on to blotting paper.

❷ When the paper is dry, draw lines with a felt-tip pen to see the affect of the size.

Make your own paper

You will need

- **an adult to help**
- paper to recycle (not too glossy)
- wire coathanger (or several)
- washing-up bowl
- scissors
- old pair of tights
- 2 large buckets
- lots of water
- food processor
- elastic band
- blunt knife

1 Tear lots of paper into small pieces and soak them in a bucket of cold water overnight.

2 Pull open a coathanger to make a squarish shape as shown.

3 Push the coathanger into one leg of a pair of tights. Tie a knot in the toe end and use an elastic band to secure the hook end. Cut off the surplus material.

4 Put a little of the soaked paper and plenty of water into the food processor and ask an adult to help you blend it for a few seconds. You should still be able to see tiny pieces of paper or fibre. Make a whole bucket of pulp this way.

5 Fill the washing-up bowl with pulp and slide your coathanger into it as shown. Be sure to keep the coathanger flat as you lift it out. Be careful not to touch it or drip water on to it.

6 Leave the coathanger flat for five minutes, then hang it up to dry.

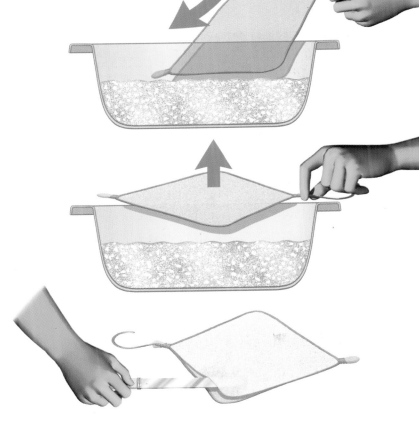

7 After a few hours, your paper will be dry. Use a blunt knife to lift it carefully from the coathanger.

8 You can add food colouring or powder paint to your pulp, or include dried petals, grasses, confetti or leaves.

Metal matters

Most of the 80 metals on Earth are good conductors. This means that heat and electricity can pass through them easily. Many of them can be shaped by beating, pulling or melting. They are shiny when cut. Some metals, such as gold, do not react easily with other substances. This means that they do not tarnish and are good for making coins and jewellery.

EXPERIMENT | Metals and magnets

You will need
- drink and food cans
- magnet

Have you ever wondered how recycling plants sort steel and aluminium cans? Both are metal, but only steel is magnetic. A huge magnet lowered on to the cans does the job in seconds.

❶ Find out if your cans are aluminium or steel by holding a magnet to them. If it is attracted, they are steel.
❷ Use your magnet to find out what else in your home is made of iron, steel, nickel or cobalt, the magnetic metals.

EXPERIMENT | A question of corrosion

You will need
- steel can
- aluminium can
- stainless steel fork
- coin

❶ Put all the items outside in a safe place.
❷ After a couple of weeks, look at the items again. Which are made of metals that react with oxygen to produce rust? Which are less reactive and suitable for exposure to liquids or use over many years?

Make a compass

You will need
- metal paperclip
- enamel paint or nail varnish
- sticky tape
- piece of cork
- magnet
- bowl of water

❶ Straighten a paperclip and mark one end of it with a dab of paint or nail varnish.

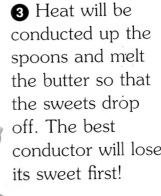

❷ Use tape to fix the wire to a cork.

❸ Stroke the wire with the south pole of a magnet from the unmarked end to the marked end about 50 times, making sure you lift the magnet high in the air between each stroke.

❹ Float the cork on a bowl of water. The wire is now magnetized, and the marked end will swing round to point north.

Good conductors

You will need
- plastic, metal and wooden spoons
- butter
- bowl of hot water
- coloured sweets

❶ Stand the spoons in the hot water with their handles resting on the bowl edge.

❷ Use a knob of butter to stick one sweet to the top of each spoon.

❸ Heat will be conducted up the spoons and melt the butter so that the sweets drop off. The best conductor will lose its sweet first!

Find out which of the materials the spoons are made of is the best conductor. If possible include spoons made of various metals in your experiment.

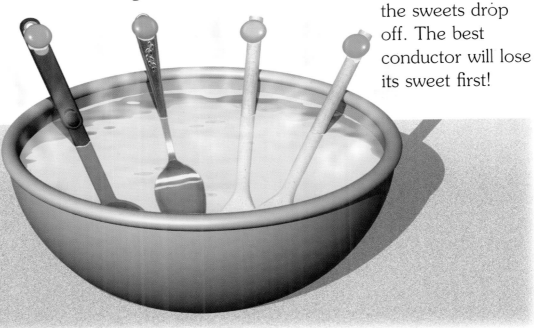

Acid or alkali?

Lemon juice is an acid, but it is very weak compared with acids used in industry, such as sulphuric acid, which can burn your skin. Alkalis are the opposite of acids. Soaps and many cleaners are akalis. Acids and alkalis can be measured on the pH scale. Below seven, the substance is acid. Above seven, it is alkali. Substances measuring exactly seven are neutral.

EXPERIMENT Make an indicator

You will need
- **an adult to help**
- half a red cabbage
- water
- saucepan
- colander or sieve
- blotting paper
- scissors
- clothes-pegs
- lemon or lemon juice
- soap

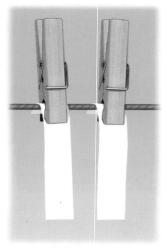

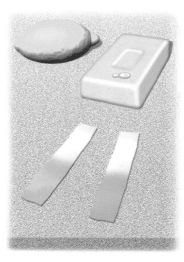

1 An indicator is a chemical that changes colour when an acid or alkali is added to it. Scientists use an indicator called litmus, but a substance found in red cabbage works well, too. Ask an adult to cut up a cabbage and boil it in a pan of water for about seven minutes. Then he or she should strain off the cabbage and leave the liquid to cool.

2 Dip strips of blotting paper into the liquid so that they soak it up.
3 Hang the strips up to dry.

4 Dip the ends of separate strips into lemon juice and soapy liquid. How do acids and alkalis change the strips? You could test other ordinary household substances, too.

Test for acid rain

You will need
- indicator strips like the ones from the last experiment
- clean jam-jar
- funnel

❶ Position a jam-jar and funnel in the open as shown. Make sure they are not under trees or buildings.

❷ When you have collected some rain in the jam-jar, test it with one of your indicator strips. If it turns red, you have collected acid rain.

When gases such as sulphur dioxide and nitrous oxide are given off by fuels and factories, they may rise into the air and combine with water vapour to form acids. They can then fall to earth again as acid rain.

Aiming for neutral

You will need
- indicator strips
- clean jam-jar
- vinegar
- soap powder
- teaspoon

If a truck overturns and spills acid or alkali, emergency teams work to neutralize the substance before it can damage people or wildlife.

❶ Put some vinegar into a jar and test it with an indicator strip.
❷ Add some soap powder to the vinegar (stirring it around to help it dissolve) and test again. What has happened to the colour on the indicator strip?
❸ Depending on how much soap powder you added, the mixture is now less acid or even slightly alkali. Experiment with the substances to see if you can make a neutral mixture. This should show no change on the indicator paper.

Floating

Why do some things sink while others float? As an object pushes water aside, the water pushes back. If the object is denser than water, the water cannot push back strongly enough to support the object, but even an object made of a very dense material may float if it contains air, as this reduces the overall density. Find out below which part of a lemon contains air!

EXPERIMENT   Sink or float?

You will need
- large bowl of water
- paperclip
- piece of fruit
- cork
- pencil
- coin
- marble

❶ Take a collection of solid objects (not hollow). The ones listed are suggestions. Write down your ideas about which items will float and which will sink.

❷ Now test your theories! Put each object into water. What happens? Note your results. The things that float are less dense than water. Those that sink are more dense.

EXPERIMENT   Do lemons float?

You will need
- lemon
- bowl of water

❶ Put a whole lemon into water. What happens?

❷ Peel the lemon and try again. Why does this happen?

Water weight

1 Tie string around a stone and weigh it on a balance. Make a note of the weight.
2 Drop the stone into water and weigh it again.

The stone is denser than the water, so it will sink to the bottom if released. But the water is still giving the stone some support by pushing back against it. This is why it weighs less in water.

Equal weight

1 Weigh an apple. Then stand a bowl in a baking pan and fill the bowl with water to the brim. Keep adding water until not another drop will go in (but none has spilled over the edge).
2 Carefully put the apple into the water. It will float, but some water will spill over the sides of the bowl.
3 Take out the apple and remove the bowl. Tip the water in the pan into the bowl of the scales and weigh it. It should weigh very nearly the same as the apple.

As the weight of water that the apple displaces is equal to its own weight, the force of the water pushing up on the apple is the same as the force of the apple's weight pushing down. The apple floats.

Dealing with density

Mass is a measure of how much matter is in a substance, but density measures how closely this matter is packed together. The density of materials is one of the things that decides how they behave together. Ice cubes float in water, for example, because frozen water contains air and so is less dense than liquid water. The experiments here show some effects of density.

EXPERIMENT Floating and sinking

You will need
- 2 drinking glasses
- warm water
- salt
- spoon
- whole, raw egg

Dissolving salt in water increases the water's density. A raw egg is denser than ordinary water but less dense than salty water. Now you know why it is easy to float in the very salty Red Sea!

❶ Fill one glass with water and carefully place an egg in it. What happens?

❷ Fill another glass with water and mix as many spoons of salt as you can into it. Now put the egg into this glass. What happens now?

Make a hydrometer

You will need

- plastic drinking straw
- waterproof felt-tip pen
- ruler
- water
- modelling clay
- salt
- spoon
- glasses or plastic containers

You could use the glasses of plain and salty water from the previous experiment if you like.

❶ Use a ruler to mark a scale on a straw at 5mm (¼ in) intervals.

❷ Place a blob of modelling clay on the base of the straw as shown.

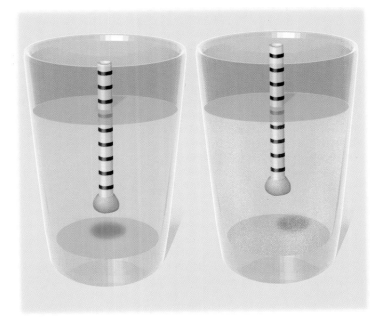

❸ Try floating the straw in plain and salty water. You can use the scale to compare the different densities.

An instrument for measuring the density of liquids is called a hydrometer. You could try measuring other liquids, such as milk, oil and juice.

Liquid layers

You will need

- cooking oil
- water
- clear plastic container with lid
- cork
- a few frozen peas

❶ Pour some cooking oil into a clear container. Then add some water. Watch what happens.

❷ Put the lid on the container and shake vigorously. Leave it to stand and watch what happens again.

❸ Open the container and put in the cork and the frozen peas. Look carefully to see where they float.

The water is denser than the oil, so it sinks. The peas are not as dense as the water but denser than the oil.

Ice magic

Why does ice, which is made of water, float in water? An enormous iceberg floats with nine-tenths of its mass beneath the water and one tenth above. It must be less dense than water, but how can that be? The experiments on this page will help you to find out. You can also discover an effect that pressure can have on temperature – and temperature can have on pressure!

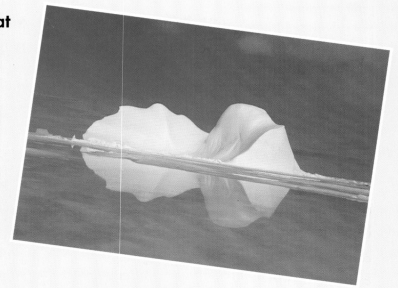

EXPERIMENT Melting moments

You will need
- glass
- ice cubes
- jug
- water

❶ Place the ice cubes in the glass.
❷ Fill the glass with water, right to the top, so that not one more drop can go in. The ice cubes will float so that parts of them are out of the water. Leave the glass. What will happen when the ice cubes melt?

The floating ice cubes displace (push aside) their own weight of water. In fact, they are made of water themselves, but with lots of tiny air bubbles trapped with it. It is these that make the ice cubes float. When the cubes melt, the air escapes, and the water left weighs exactly as much as the water it has displaced. So the glass does not overflow!

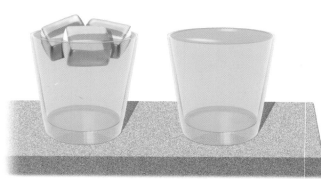

Trapped in the ice

How can you thread a wire through an ice cube without drilling a hole? Try using the power of melting and freezing!

1 Fill two plastic bottles with water and join them with a short length of very thin wire or fishing line tied around their necks.

2 Put an ice cube on a saucer on top of a pile of books and balance the bottles as shown so that they hang on either side.

The weight of the bottles on the wire puts great pressure on the ice. This raises the temperature so that the ice melts and the wire sinks into it. Behind the wire, as the temperature drops again, the ice freezes up.

Growing colder

2 Put the bottle in a plastic bag and stand it upright in the freezer. Wait a few hours.

1 Fill a bottle to the brim with water and push the cork in just a little way (not too tightly).

When you take the bottle out, the water will have frozen. As water expands when it freezes, the top will have popped off! The plastic bag is there just in case the ice shatters the glass. If so, don't touch it. Ask an adult to help you dispose of it safely.

Amazing mixtures

Most materials that we use are not made of only one kind of molecule. The simplest way in which molecules can exist together is by mixing, which does not change the molecules themselves. If you mix sugar and sand, for example, each tiny grain is either sugar or sand, not a combination. Simple mixing sometimes means that there is a fairly simple way of unmixing!

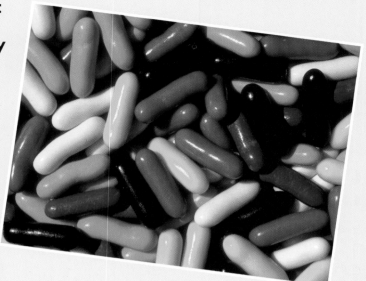

EXPERIMENT Mixing the unmixable

You will need
- glass or plastic jar
- cooking oil
- water
- spoon
- washing-up liquid

Oil and water don't mix. You may have seen pictures of oil spillages at sea, where the oil floats thickly on the water. But science lets us mix the unmixable.

❶ Pour some water and some oil into a jar. What happens?
❷ Vigorously mix them with a spoon and wait. What happens now?
❸ Now add a few drops of washing-up liquid and mix again. What do you find?

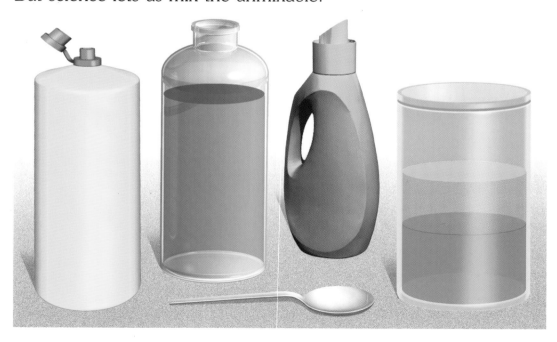

The washing-up liquid acts as an emulsifier, breaking the oil into tiny droplets that are dispersed among the water molecules.

26

EXPERIMENT Unmixing

You will need
- fine table salt
- water
- large flat pan or baking tray
- spoon
- bowl or jug

❶ Pour some water into a bowl or jug and mix in as much salt as you can, stirring in a little at a time until it is dissolved.

❷ Pour the salty water into the flat pan and leave it in a warm, airy place for a few days. What do you find?

In time, the water evaporates (turns into vapour) and rises into the air. What you have left is the salt! Evaporation is a useful method of separating some mixtures and has been used for centuries to extract salt from sea water.

EXPERIMENT Unmixing another way

You will need
- water
- caster sugar
- plain flour
- spoon
- jar or glass
- jug
- funnel
- coffee filter

❶ In a dry jug, mix some sugar and flour together. The grains are tiny – how could you ever separate them?

❷ Fill the jug with water and mix well. Dip a finger into the water and taste it. The flour sinks to the bottom, but the sugar dissolves into the water.

❸ Put the filter paper into the funnel and stand it in a glass or jar. Gradually pour the mixture into it. Finally, pour a little hot water through.

The sugar and water molecules are small enough to pass through the filter, but the flour ones are not. If you taste a tiny bit of the flour, you should find that the sugar has gone.

Think about the experiment above. How could you now get back the flour and sugar, if you didn't mind losing the water? Try it!

Chemical compounds

As well as simple mixtures, there are more complicated ways in which molecules can form new substances. Atoms in the molecules can react chemically with each other, forming new molecules that contain atoms from two or more original molecules. Substances formed in this way are called compounds. Their atoms are held together strongly and are difficult to separate.

EXPERIMENT Reasons for rust

You will need
- iron filings
- water
- small glass jar or bottle
- saucer

❶ Rinse the inside of a bottle with water and tip some iron filings inside. Shake them about so they stick to the sides.

❷ Half fill a saucer with water and balance the bottle upside down in it. Leave it overnight.

In the morning you will find that the filings have rusted and the water has risen a little way up the jar. That is because the filings have reacted with oxygen in the air to form a compound – iron oxide. The water moves into the space left by the oxygen.

28

EXPERIMENT Compound change

You will need
- drinking glass
- vinegar
- sodium bicarbonate (sometimes called bicarbonate of soda)

1 Shake a little sodium bicarbonate into a glass.

2 Pour in some vinegar. The mixture will at once begin to fizz, which means that a gas is being given off. What could it be?

Sodium bicarbonate is a compound that is changed by coming into contact with the vinegar. Its elements are sodium and carbon, and the gases hydrogen and oxygen. The vinegar causes the carbon and oxygen to form a compound called carbon dioxide, which is a gas – the same one that makes your drinks fizzy. Something that causes a chemical change, like the vinegar, is called a catalyst.

EXPERIMENT Compound fracture

You will need
- 2 small, clean spice jars
- 3 volt battery
- water
- salt
- cooking foil
- 2 plastic-covered copper wires
- glass bowl
- insulating tape
- piece of pencil lead

1 Wrap one exposed end of one wire around the pencil lead and cover the join with tape. Attach the other end to the positive (+) terminal of the battery. Cover one exposed end of the next wire with foil and attach its other end to the negative (–).

2 Fill a bowl with water and mix in as much salt as you can. Fill two spice jars with the solution and turn them upside down in it. Arrange the wires as shown.

Salt is a compound of sodium and chlorine. Water is a compound of hydrogen and oxygen. Electricity breaks these down, producing hydrogen and chlorine gas, which bubble up into the jars.

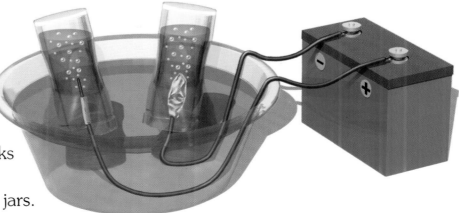

Food science

All cooks are scientists. They know that putting different substances together produces new mixtures. Heating substances together can create new compounds – although we may think of them as sweets and savouries! Our foods are made of chemical substances just as we are. Cooking is all about making chemical changes – hopefully delicious ones!

EXPERIMENT Starch or not?

You will need
- iodine
- flour
- items of food
- old, clean jam-jars
- water
- plastic teaspoons
- rubber gloves
- biological washing powder
- knife

You can buy iodine (iodine solution or tincture of iodine) from a chemist, but be very careful. It can stain your skin and anything else it touches.

❶ Wearing rubber gloves, put a little flour into a clean jam-jar and add some water to make a very runny liquid. Stir in a drop or two of iodine solution. The liquid will turn deep blue, showing that there is starch present.

❷ Add a little biological washing powder to the mixture. The powder contains enzymes that destroy the starch. What happens to the colour of the liquid?

❸ Now test small amounts of other foods. Powders and liquids can be mixed with water. Solid items, such as raw fruit and vegetables, can be painted with a little iodine on a cut surface.

Our bodies need carbohydrates, which are starches, to be healthy. You should be able to discover which kinds of foods contain them. You only need a very small amount to test but you must throw it away afterwards.

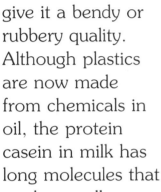

EXPERIMENT Produce your own plastic

You will need
- **an adult to help**
- milk
- vinegar
- small saucepan
- hob

Plastic is made of long molecules that give it a bendy or rubbery quality. Although plastics are now made from chemicals in oil, the protein casein in milk has long molecules that work as well.

1 Ask an adult to help you heat a small amount of milk in a saucepan. It needs to be warm but not boiling or foaming.

2 Pour in a little vinegar. You will find a rubbery substance in the pan. Wash it carefully under the tap, being careful of the hot pan.

3 You can shape the rubbery stuff, which is a form of plastic, into flat shapes, which will harden within a few days.

EXPERIMENT (Almost) instant ice cream

You will need
- thick milkshake
- clean yogurt carton
- ice
- plastic bowl
- strong plastic bag
- rolling pin
- salt
- chopping board

This is how ice cream used to be made! Salt lowers the freezing point of the ice, and the ice cream freezes.

1 Half fill a yogurt pot with thick milkshake from the fridge. Cover the pot with a lid or tie it in a plastic bag.

2 Put lots of ice in a plastic bag and crush it with a rolling pin.

3 Put the ice in a bowl and quickly mix in a third as much salt. Push the yogurt pot well down into the ice. Put a little more ice over the top and cover the bowl with a chopping board. Wait 15–20 minutes.

Strong structures

Just as important as the properties of materials used in building is the way in which they are used. Some structures are very much stronger than others. Architects and engineers need to be aware of the loads and stresses on buildings, including such things as changes in temperature, weather conditions and use. These experiments help you to investigate further.

EXPERIMENT Bridging the gap

You will need
- paper or thin card
- books
- yogurt pot
- marbles

It makes sense to use strong materials to build a bridge, but structure is just as important. Try the following structures to see which is strongest.

❶ Build two level "banks" of books and lay a single sheet of paper or thin card across it. Balance an empty yogurt pot on top and test the strength of your bridge by seeing how many marbles you can drop into the pot before the bridge collapses.

❷ Use the same piece of paper or card but this time fold it like a fan, as shown. How many marbles can it support now?

❸ This time, make an arch to support your first bridge. Simply bend paper or card as shown. Lay the single piece of paper or card over the top. Is it stronger than **❶**?

Tower power

1 Take one sheet of paper and fold it in half lengthwise. Open it out and fold each half into the middle. Tape the edges to make a square tower.

2 Join the long edges of another sheet of paper to make a tube. Then make another three square towers and another three tubes.

3 Set up your towers as shown, taping the bases to card. Lay books on the top to find which is the strongest structure.

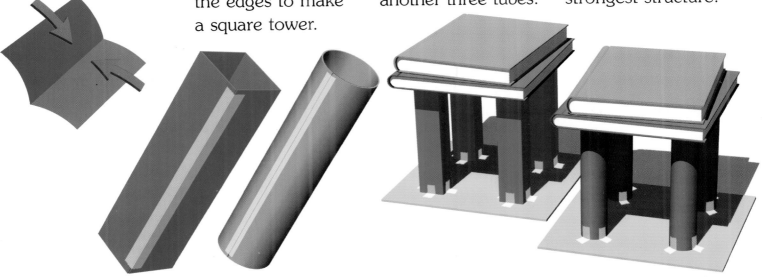

A buttressed building

You may see buttresses on old buildings, such as churches and castles. Some slope out from the wall, giving even more support.

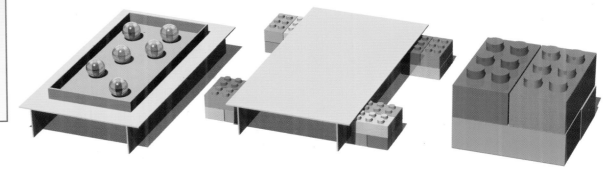

1 Make a simple building from card as shown. Attach the walls with a single strip of tape on each side.

2 Carefully put the plastic tray on top and put in marbles until the building collapses.

3 Remove the tray and build some buttresses for your structure from building bricks. You will need to make four cube-shaped structures to stand next to the walls. Now test the strength of the building again. What do you find?

Glossary

Absorbency The ability of a material to soak up liquid.

Acid A substance with a pH value of less than seven.

Alkali A substance with a pH value of more than seven.

Atom The smallest part of an element (substance made up of only one kind of atom) that can exist.

Catalyst Something that causes a chemical change to take place.

Conductor A material through which heat and electricity can pass easily.

Convection The movement of heat through a liquid or gas by means of a current.

Corrosion The reaction of a metal with oxygen in water or the air to produce rusting.

Density A measure of how tightly the matter in a substance is packed together.

Displacement The way in which an object immersed in a liquid pushes away that liquid.

Emulsifier Substance that breaks liquids such as oil into tiny droplets.

Evaporation The turning of a liquid into a vapour, or gas.

Fabric A material made from fibres that have been woven, knitted or pressed together.

Grain The way in which the fibres in paper tend to line up and lie in the same direction.

Hydrometer An instrument for measuring the density of liquids.

Indicator A chemical that changes colour when an acid or alkali is added to it.

Mass A measure of how much matter is in a substance.

Molecule The smallest particle of a substance that can exist by itself.

Neutral A substance that is neither acid nor alkali, with a pH of seven.

Siphon A tube that makes use of air pressure to move water from a higher to a lower level.

Size A substance added to paper pulp or to the surface of paper to make it less absorbent.

Water pressure The force with which water presses down on something beneath or within it.

Living
with Light

Light for growth

Plants can do something that no other living thing can do. They can convert energy from the Sun into food energy to help them live and grow. Of course, plants need other things too, such as water and nutrients from the soil. Without plants, there could be no animal life on Earth. Animals that eat plants are in turn food for animals that eat meat.

EXPERIMENT What do plants need?

You will need
- cotton wool
- cress seeds
- 4 saucers
- water

❶ Put a layer of cotton wool on each saucer. Sprinkle some cress seeds on top.

❷ Sprinkle the first saucer with plenty of water and put it on a sunny windowsill.

❸ Do the same with the second saucer but put it in a very dark cupboard.

❹ Put the other two saucers on the windowsill and in the cupboard as they are.

❺ After a few days, look at the saucers again. The plants should be at very different stages. What does this tell you about the conditions that plants need?

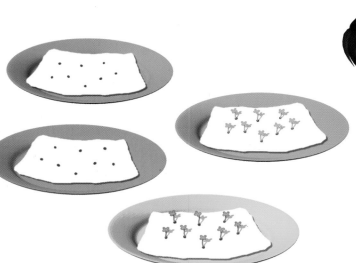 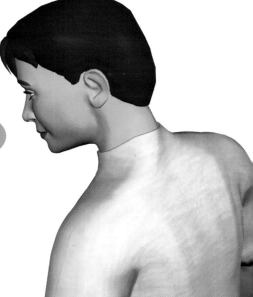

How do plants use light?

You will need
- a pot-plant with broad green leaves, such as a geranium
- a round sticking plaster

1 Stick the plaster on to a leaf and put the plant in a sunny place. Don't forget to water it!

2 After a week, carefully peel off the plaster. The leaf should be paler underneath. The green part of plants contains chlorophyll, which is able to convert light energy into food energy. This is called photosynthesis.

EXPERIMENT # Searching for the Sun

You will need
- small potato
- shoebox with lid
- card
- scissors
- sticky tape

1 Cut a small hole in one end of the shoebox as shown.
2 Use pieces of card to make partitions in the box. They should only stretch part of the way across. Together, they will make a kind of maze.
3 Put the potato at the opposite end from the hole and tape the lid on to the box. Place the box on a sunny windowsill.

4 After several weeks, a shoot from the potato will find its way through the maze and into the light. Open the box to see its twisting path. Plants have a property called phototropism, meaning that they grow towards light, even if they are deep in the soil.

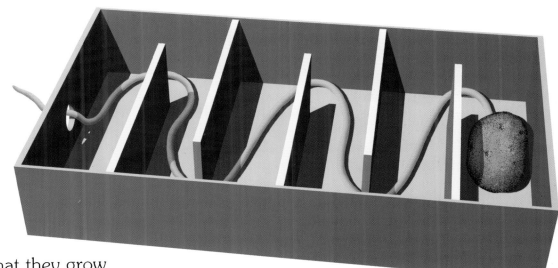

Light on the move

Light travelling from an object strikes our eyes and enables us to see the object. Normally, light travels in a straight line, but if it moves through different materials, such as glass or water, its speed is changed and its path appears to bend. This is why a pool of water seems shallower than it really is when viewed from above.

EXPERIMENT How does light travel?

You will need
- card
- scissors
- modelling clay
- knitting needle
- hole punch
- torch

1 Do this experiment when it is dark or cloudy outside. Cut some card into two rectangles. Use the hole punch to make a hole in the centre of each rectangle. Ask an adult to help you if necessary.

2 Push a knitting needle through the holes to line them up exactly. Then use modelling clay to fix the cards upright on a flat surface.

3 Remove the knitting needle. Turn off the light and shine a torch through the holes. Because light travels in a straight line, it will pass through in a narrow beam.

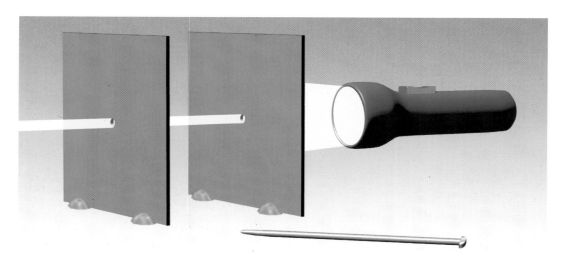

4 Remove the cards and shine the torch on to a light-coloured wall. Try making shadow shapes with your hands. Shadows are made because the light travels in straight lines. It cannot go around an obstacle, so a dark place is left on the far side.

What is refraction?

- drinking glass
- water
- pencils

1 Light always tries to travel in a straight line, but moving from one transparent substance to another, such as from air to water or glass, can change its speed and cause it to bend. Put some pencils in a glass of water. Viewed from the side, they seem to bend where they enter the water.

Bending beams

You will need
- drinking glass
- coin
- water

1 Drop a coin into the bottom of a glass and place your hand in front of it.

2 Pour water into the glass. You will see the coin appearing to float on top of it.

3 Remove your hand. The coin will seem to be both floating and lying at the bottom.

It is the light rays coming from the coin that you see. The water and glass make them bend so that you seem to see the coin in two places at once.

Crossing lights

You will need
- a shoebox
- sheet of white paper
- scissors
- glass of water
- torch

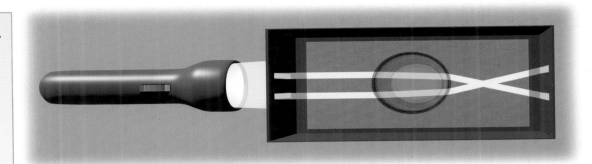

1 Cut two slits about 2.5cm (1 in) apart in one end of the box. Put the white paper in the bottom of the box and place the glass of water in the middle.

2 Turn off the light and shine the torch through the holes. What happens to the light?

Light into colour

Light from the sun or an electric light looks white or colourless, but in fact it is made up of lots of colours. These colours can be seen if a beam of light is split by a triangular lens called a prism. On bright, wet days, drops of rainwater sometimes act as prisms, splitting light into the arched band of colours we know as a rainbow.

EXPERIMENT Make a prism

You will need
- rectangular cake tin
- small mirror
- white surface or sheet of white paper
- modelling clay
- water
- ruler

❶ Fill a square cake tin with water and place it in front of a white surface.

❷ Angle a mirror in it so that it forms a triangular area of water, which will act as a prism.

❸ When the water is completely smooth, gently move the mirror until you see a rainbow on the white surface.

❹ Fix the mirror in place with a ruler and blobs of modelling clay.

EXPERIMENT Make colours disappear

You will need
- piece of white card
- scissors
- pencil
- pair of compasses
- coloured pens or paints

 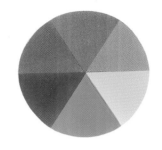 

❶ Draw a circle on white card and divide it into six segments as shown. Cut out the circle.

❷ Colour each segment in turn red, orange, yellow, green, blue and violet, in that order.

❸ Push a pencil through the middle and spin it. The colours will become a yellowish white.

EXPERIMENT Separating colours

You will need
- blue ink
- blotting paper or filter paper
- water

❶ Drop a small amount of ink on to blotting paper. The colour should spread out a little.

❷ Drip a drop of water into the centre of the colour. The spot will spread out further. Some of the dyes in the ink will travel across the paper more easily than others, so they will divide and show rings of different colours. This is called chromatography.

Painting with light

"White" light is made up of all the colours of light mixed together. Grass looks green because it absorbs all the other colours but reflects green light. Mixing coloured lights gives quite different effects from mixing the pigments that give colour to paints. Try these experiments to find out more about these differences.

EXPERIMENT Mixing pigments

You will need
- red, yellow and blue paints
- small paintbrush
- paper
- magnifying glass

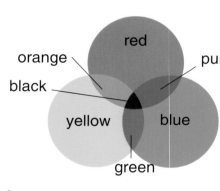

red
orange
purple
black
yellow
blue
green

❶ Remember how you learnt to mix colours when you were little? Look at the colours above to remind yourself. Towards the end of the nineteenth century, a painter called Georges Seurat tried mixing colours in a different way. He used tiny dots of pure colour placed close to each other. Try putting red and yellow dots close together. How do they look from far away? Can you make a more complicated picture or pattern using the same method?

❷ Now use a magnifying glass to look at the pictures in this book. They are all made from dots of these four colours:

42

EXPERIMENT Mixing light

You will need
- white paper
- 3 torches
- red, blue and green cellophane
- sticky tape

1 How do coloured lights combine to make other colours? Use sticky tape to cover torches with red, blue and green cellophane.

2 Darken the room and shine the torches on to white paper. What happens where the colours overlap?

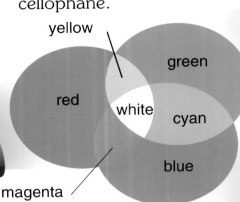

EXPERIMENT Make some 3D glasses

You will need
- old pair of sunglasses
- red and green cellophane
- card
- scissors
- sticky tape

1 Draw round some old sunglasses on to card to get the shape of your glasses. You will need a front piece and two side pieces.

2 Use sticky tape to join the pieces together. Then glue a piece of red cellophane over the left eye hole and a piece of green cellophane over the right one. If you wear your glasses to look at a special 3D photograph the picture will seem to have "depth". Each eye only sees part of the image. Your brain puts the two views together to give a three-dimensional effect.

43

Light in action

Without light, we could see nothing. It is light passing into our eyes through the lens at the front that enables our brains to give us a view of what is around us. But light can be very useful in other ways. By knowing how it passes through or is reflected by different materials, we can use light for many different purposes.

EXPERIMENT Fibre optics

You will need
- plastic bottle
- torch
- water
- knitting needle

❶ Take the top off a plastic bottle and carefully make a hole in the side with a knitting needle.

❷ Put your finger over the hole and fill the bottle with water over the sink. Put the top back on.

❸ Ask a friend to turn off the light in the room and shine a torch behind the bottle as shown.

❹ Take your finger away from the hole and put it in the stream of water now coming from the bottle. You will see light from the torch on the end of your finger.

The light beams have bounced backwards and forwards inside the bottle and the spout of water. This is how light travels along fibre optic cables.

EXPERIMENT Make a mini-microscope

You will need
- small mirror
- modelling clay
- drinking glass
- aluminium foil
- needle
- sticky tape
- drop of water
- small flower

❶ Microscopes use a glass lens to bend or refract light. A drop of water can act as a lens. Angle a mirror on a blob of modelling clay and place a glass on top.

❷ Take a piece of aluminium foil and fold it like a fan to make a strip of several layers. Pierce the centre carefully with a needle, jiggling it to make a small hole.

❸ Bend the foil over the glass as shown, using sticky tape to secure the edges. With your fingertip or the needle, drip a small drop of water into the hole in the top.

❹ Place a small flower or another tiny object under the water-drop lens on top of the glass. Your homemade mini-microscope will magnify it up to 50 times.

EXPERIMENT Light, heat and colour

You will need
- black paper
- white paper
- elastic bands
- 2 identical jam-jars or glasses
- water
- measuring jug
- thermometer

❶ Fix a piece of white paper around one glass or jar using one or more elastic bands. Fix a piece of black paper around the other.

❷ Carefully fill the containers with water, making sure they contain the same amount.

❸ Leave the containers on a sunny windowsill for an hour or two, then measure the temperature in each.

Dark surfaces absorb more heat and light, while light surfaces reflect it. That is why lighter clothes keep us cooler in summer.

Bouncing light

Knowing about the way that light travels can be very useful. The machines on these pages all use the fact that light acts in a predictable way. A periscope enables sailors in a submarine to see above the surface. A large version of a pinhole camera, called a camera obscura (darkened room) was in use as long ago as the fifteenth century.

EXPERIMENT Make a periscope

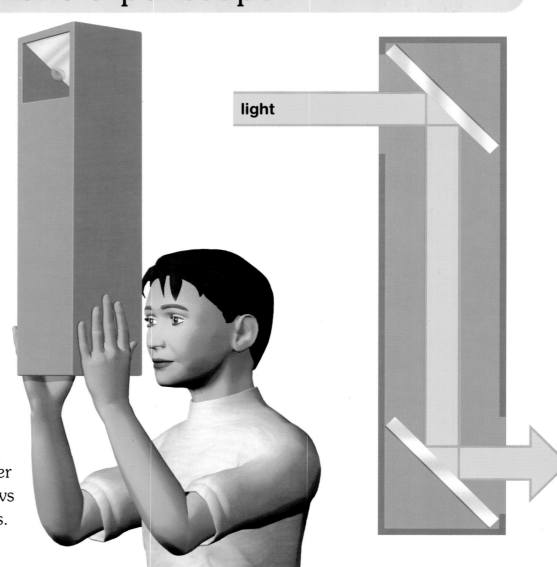

You will need
- a long, thin cardboard box
- scissors
- 2 flat mirrors
- sticky tape
- modelling clay
- protractor for measuring angles

❶ Cut squares out of the top and bottom of a long, thin box as shown.

❷ Use tape and modelling clay to position two mirrors inside the box at exactly 45°.

❸ Now you can see over walls! The diagram shows how the periscope works.

light

EXPERIMENT — Make a kaleidoscope

You will need
- 3 mirrors of the same size
- white paper
- sticky tape
- coloured paper shapes

1 Tape three mirrors together so they form a three-sided tube. Position them on white paper.

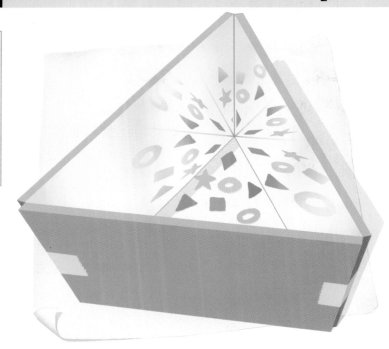

2 Drop some coloured paper shapes into the tube and look at the patterns they make. What happens if you move the shapes with a pencil?

EXPERIMENT — Making a pinhole camera

You will need
- small box
- sticky tape
- scissors
- card
- tracing paper
- pin

1 Cut away one side of a small box and use tape to fix some tracing paper over it. Make sure it is taut.

2 Make a tiny hole with a pin in the centre of the opposite side.

3 Point the pinhole towards a bright window. You will see an image of the window on the tracing paper screen. Move backwards or forwards until the image is clear.

4 Did you notice that the image was upside down? You can see why in the diagram below. The same thing happens when light enters your eye, but your brain turns everything you see the right way up again for you!

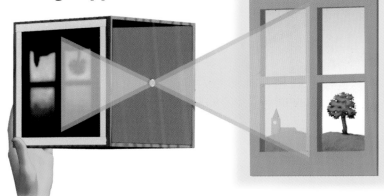

Light from the Sun

Thousands of years ago, people noticed how shadows moved throughout the day and realized that they could use this as a way of telling the time. Sundials work because light travels in straight lines. But sunlight also hits tiny particles of dust in the atmosphere as it travels towards us. This changes the way it appears at different times of the day.

EXPERIMENT Make a sundial

You will need
- **an adult to help**
- flat piece of wood about 50cm (20 in) square
- 50cm (20 in) length of dowel
- wood glue
- waterproof paints
- paintbrushes

❶ Ask an adult to make a hole at one edge of a flat piece of wood and help you fix a length of dowel into it, using glue if necessary.

❷ Decorate the sundial with waterproof paints. Place it outside on a sunny day.

❸ Once an hour, mark the position of the shadow on the base and write or paint the time next to it.

Remember: to use your sundial again, you must always place it in exactly the same position.

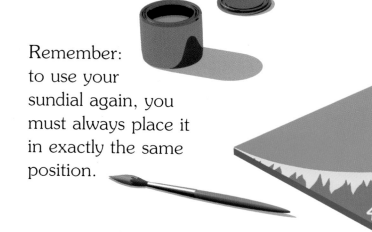

The path of the Sun

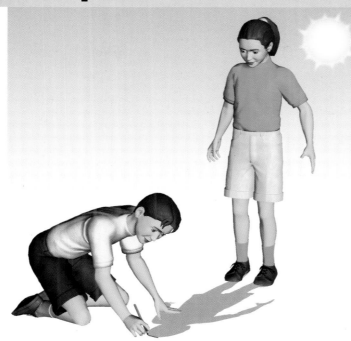

You will need
- a friend to help you
- chalk
- flat, firm surface, such as a playground

1 On a sunny day, ask a friend to draw around your shadow with chalk on the ground.

2 At hourly intervals, repeat step **1**. Make sure you stand in the same place and face the same way each time. Not only does your shadow move around, it also shortens and lengthens as the Sun moves across the sky.

What colour is the sky?

You will need
- drinking glass
- water
- teaspoon
- plain flour
- white paper or card
- torch

1 Mix half a teaspoon of flour into a glass of water.
2 Put the glass on a piece of white paper and shine a torch on it from above. The water should look very slightly blue or grey.

3 Now put the paper behind the glass and shine the torch at it sideways. The water will look very slightly orange or yellowish.

Tiny particles in the air, like the flour in the water, scatter the colours in light. When light shines from the side (or the Sun is low in the sky), blue light scatters and your eyes see more orange light.

Natural light and colour

The natural world is full of light and colour. No one knows exactly what the world looks like to an insect, but we can do experiments to show if colour makes a difference to it. Natural dyes from plants, animals and minerals have been used for centuries. Try them yourself. You can do an experiment to create lightning in your own home, too!

EXPERIMENT Insect favourites

You will need
- white and pink cardboard
- scissors
- black and red felt-tip pens
- wire (you could ask a grown-up to cut some from old coathangers)
- modelling clay
- honey

❶ Cut out two pink and two white flower shapes like the ones below. Draw red "honey guides" on the two pink flowers. Draw black "honey guides" on one of the white flowers.

❷ Bend wires to make stems and push them through the centres of the flowers, using modelling clay to keep them in place. Push the wires into the ground outside near some real flowers.

❸ Put a blob of honey in the centre of the all-white flower and one of the pink flowers.

❹ Watch carefully from a metre (yard) or so away. Are insects more attracted to the white or coloured flowers? Try the experiment again, testing different colours.

Vegetable dyes

You will need
- **an adult to help**
- small saucepan
- water
- onion skins
- white cotton wool
- spoon

1 Ask an adult to boil lots of brown onion skins in a little water for 15 minutes.

2 Allow the water to cool, then put some cotton wool in the liquid. Leave it to soak for five minutes, then fish it out with a spoon and leave it to dry. What colour is it?

3 You could try the same experiment with beetroot, spinach or tea leaves.

Make your own lightning

You will need
- metal baking tray
- modelling clay
- plastic carrier bag
- metal fork

1 Take a fistful of modelling clay and press it on to the bottom of a baking tray to make a kind of handle. From now on, touch only the modelling clay, not the tray itself.

2 Holding the tray by the clay "handle", rub it round and round on a plastic carrier bag flat on a table. This builds up static electricity on the tray. Don't let the tray move off the bag.

3 Lift the tray (still using the clay handle) a short distance from the plastic and bring the prongs of a fork towards one of its corners. A spark of "lightning" will jump from the tray to the fork, just as lightning jumps from a cloud to a lightning conductor.

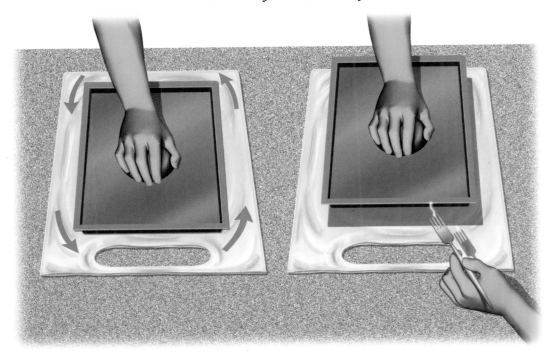

Moving pictures

To make sense of what we see, we need more than good eyesight. We need a brain to interpret the images that flash into our eyes. But the brain, powerful as it is, can only deal with a certain number of images per second. More than this, and it "fills in the gaps" for us, making still images appear as moving pictures.

EXPERIMENT Make a flicker book

You will need
- small sheets of paper or thin card
- felt-tip pens
- tracing paper
- pencil
- stapler

❶ Draw a picture on a piece of paper, leaving a margin on one side. The picture should have one object in it that can move. The ideas below could give you some ideas.

❷ Trace the main lines of the picture on a fresh sheet of paper and complete the picture, but make sure that the moving object changes position a little bit.

❸ Repeat the last step until you have enough pages. You will need at least 12 for a good effect – more is better!

❹ Put the sheets in a pile. Staple the edge. Flick the pages to see your picture "move".

An even easier flicker book!

You will need
- computer with a drawing program
- printer
- stapler as before

1 Today cartoon-makers (called animators) use computers. You can easily follow their example. First create a picture on your computer and save it.

2 Next make a small change to the picture and save it under a different name. The easiest thing is to add 1, 2, 3 and so on to the file name as you go along.

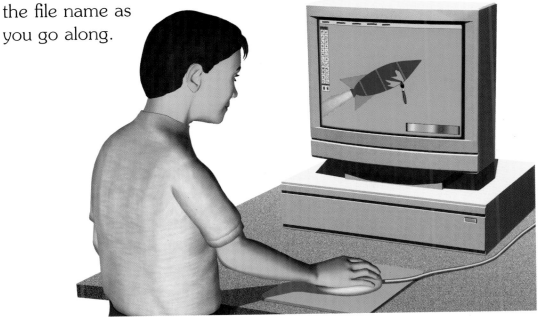

3 Repeat the previous stage until you have finished, then print out your pictures and staple them together as before. Using this method, it is easy to create effects such as darkness falling, or the leaves on a tree gradually changing colour.

Over and over

You will need
- small piece of card
- felt-tip pens
- scissors
- 2 elastic bands
- coathanger

1 On one side of a small piece of card draw a cartoon prison cell with thick bars.

2 On the other side of the card draw the face of a teacher, friend or enemy!

3 Make a small hole at the top and bottom of the card. Loop an elastic band through each hole.

4 Tie the elastic bands to the top and bottom of a coat-hanger so that the bands are quite taut.

5 Turn the card around and around – then let go! Messages about the two images enter your brain too quickly to make sense, so it makes them into one picture.

Glossary

Animator A person who makes moving cartoons by creating a series of images with slight differences from one to the next. These are then filmed and shown quickly one after the other.

Chromatography Separating the colours in inks and other liquids by allowing them to spread across absorbent paper.

Dyes A substance (usually used in liquid form) that can change the colour of fabric, hair, skin and so on.

Fibre optics Fine glass fibres that can carry light.

Honey guides Markings on the petals of flowers that guide insects towards the centre of the flower where pollen and nectar are to be found.

Kaleidoscope A tube with mirrors that can make patterns out of tiny shapes placed in the middle. The patterns change if the kaleidoscope is shaken or turned.

Microscope A machine using a lens or several lenses to bend light so that tiny objects appear bigger.

Nutrients Substances that living things need to take into their bodies to stay alive and healthy.

Periscope A machine using mirrors to bounce light around corners, making it possible to see things that are otherwise hidden.

Photosynthesis The way in which plants are able to make food energy out of light energy from the Sun.

Phototropism A plant's ability to grow towards the light, even deep underground.

Pigment A substance that can give colour to another substance.

Prism A block of glass or other material that splits light into the colours of the rainbow when it passes through.

Rainbow An arched, coloured band of light in the sky caused when droplets of rain or spray act as prisms to split sunlight. Traditionally, rainbows are said to have seven colours: red, orange, yellow, green, blue, indigo and violet.

Refraction The way that the speed of light changes and the light bends when it passes from one substance to another, such as from air to water or through a lens.

Movement and Machines

Marvellous motion

Sir Isaac Newton stated three laws of motion. The first says that an object will stay still or continue to move at the same speed and in the same direction unless it is acted on by a force. The second says that when a force acts on an object, it will cause it to move, change direction, speed up or slow down. The third states that every action produces an equal and opposite reaction.

EXPERIMENT Table trick

You will need
- plastic cups
- water
- tablecloth
- table outdoors

❶ Spread the cloth on the table. Fill the cups with water to make them heavier and stand them on the cloth.
❷ Take hold of the edge of the cloth and give it a hard, sharp tug. You should be able to pull the cloth away without disturbing the cups.

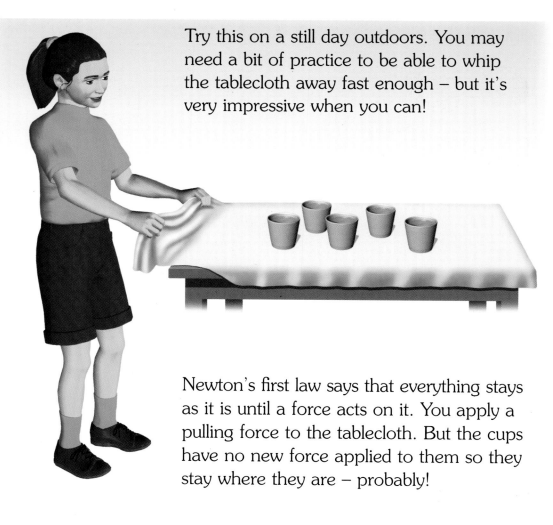

Try this on a still day outdoors. You may need a bit of practice to be able to whip the tablecloth away fast enough – but it's very impressive when you can!

Newton's first law says that everything stays as it is until a force acts on it. You apply a pulling force to the tablecloth. But the cups have no new force applied to them so they stay where they are – probably!

EXPERIMENT Parachute pulls

You will need
- cotton handkerchief
- fine string or ribbon
- clothes-peg

1 Tie two pieces of string or ribbon of equal lengths to the diagonal corners of the handkerchief as shown.
2 Clip a clothes-peg over the place where the two strings cross.

3 Hold the parachute by the middle of the handkerchief and hold it up high. Let go. What happens? Two forces are acting on the parachute: gravity and the force of air resistance. As gravity is the stronger force, it gradually pulls the parachute down to the ground.

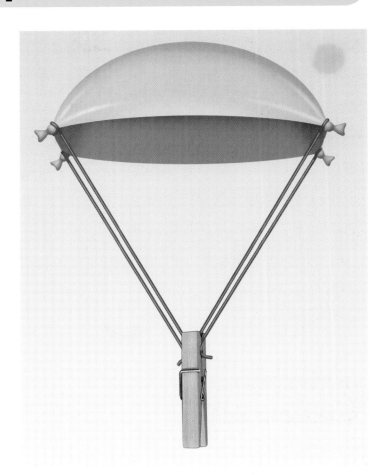

EXPERIMENT Actions and reactions

You will need
- rollerskates, in-line skates or skateboard
- beach ball
- friend to catch you!

1 Position yourself in the middle of an open area on a flat, firm surface, wearing skates or standing on a skateboard and holding a ball.
2 Throw the ball as hard as you can away from you. What happens to you? Which direction do you move in? What do you think would happen if you threw the ball backwards over your head? Try it!

57

Stopping and going

An object at rest has a property called inertia. It will not start moving until a force is applied to it. Forces such as gravity and friction will soon slow the object down unless more force is applied. The weight of water supplies this force in the experiment opposite. In the one below it, only the bag with the same length of string gets a useful push from the larger bag.

Spinning egg

You will need
- 1 hardboiled and 1 fresh egg
- 2 saucers

❶ Put one egg on each saucer and try to spin the eggs. What happens?
❷ Now touch each egg lightly with a fingertip to stop it spinning. What happens this time?

Two things that look identical may have very different properties when it comes to motion. Try this experiment, then show it off to your friends.

The hardboiled egg spins and stops easily because its insides are solid and fixed. Inside the raw egg, the yolk can move in the white, making its spin less stable. When you try to stop the outside, the yolk keeps moving, making the shell move again as well.

Turning toy

You will need
- **an adult to help**
- child's plastic bucket
- string
- 2 bendy drinking straws
- scissors
- jug of water
- modelling clay

❶ Ask an adult to make two holes near the base of the bucket but on opposite sides.

❷ Cut two straws about 7.5cm (3 in) from the bendy end and push these through the holes.

❸ Bend the ends so that they are parallel to the bucket and pointing in opposite directions when viewed from above. Seal around the straws with modelling clay.

❹ Tie a piece of string through the handle of the bucket. Hold the bucket by this over a sink and pour water into it. What happens? What happens when there is no more water?

EXPERIMENT Swinging in sympathy

You will need
- 2 chairs
- strong string
- 6 small plastic bags
- marbles

❶ Put three marbles in each of five bags. In the last bag, put 10 marbles.

❷ Tie string between two chair backs. Then suspend each of the bags from this string, making the strings for the bags different lengths. Make sure that the large bag and one other have the same length of string. Now start the heaviest bag swinging. What happens to the others?

Energy sources

Almost all the energy on Earth comes either directly or indirectly from the Sun. Plants are able to convert Sun energy into food energy. It is the dead bodies of plants (and animals that have eaten plants), crushed under tremendous pressure for millions of years, that become fossil fuels – coal, oil and gas. Even **your** energy comes from the Sun in the first place.

EXPERIMENT Solar-powered fan

You will need
- kitchen foil
- black paint or marker
- scissors
- sticky tape
- thread
- large, clean jam-jar with lid

❶ Cut two strips of foil about 2.5cm (1 in) by 10cm (4 in). Blacken the non-shiny side with paint or a marker. Cut slots in the strips as shown and fit them together, bending the ends down as in the picture.

❷ Use thread attached by sticky tape to fix the solar panels to the lid of the jar. Replace the lid and leave the jar in a sunny place. The sun warms the black sides of the foil more than the shiny sides, causing changes in air pressure that make the fan move.

Centrifugal lift

You will need
- cotton reel
- strong string
- scissors
- plastic bag
- marbles
- reel of sticky tape

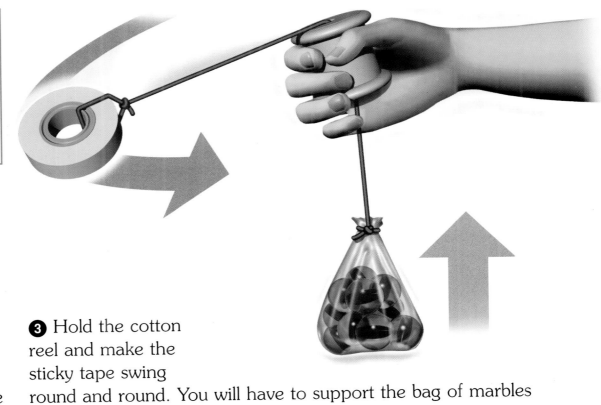

1 Cut a piece of string about 25cm (10 in) long and pass it through the cotton reel.

2 Tie a bag full of marbles on one end as shown and a reel of sticky tape on the other end.

3 Hold the cotton reel and make the sticky tape swing round and round. You will have to support the bag of marbles just at the beginning. Then let go and see what happens.

Water-powered winder

You will need
- strong cardboard
- scissors
- knitting needle
- wool
- plastic straw
- sticky tape
- modelling clay

1 Lay two pieces of cardboard beside the sink and tape pieces of drinking straw to them so that a knitting needle can slide through and turn freely.

2 Cut "blades" of card and use modelling clay to fix them to the knitting needle as shown. Use a small blob of modelling clay to fix the end of a piece of wool further along the needle.

3 Run the tap gently over the card blades and the wool should wind on to the needle.

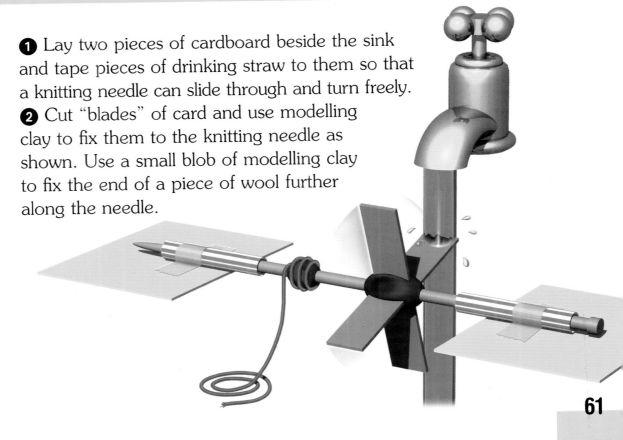

Power sources

Newton's Second Law says that when a force acts on an object it will move, change direction or slow down. Heat – even a small amount – can supply a force for movement. The warmth of your hands can cause cold air to heat up and expand in the experiment below. An elastic band can store your own energy, put into it by your muscles when you twist it.

EXPERIMENT Hot-air power

You will need
- clean, empty glass bottle with lid
- water

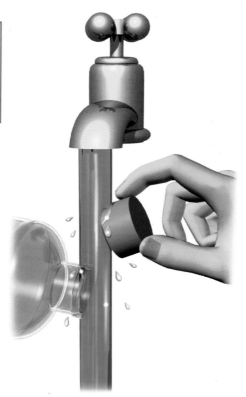

❶ This experiment will work better if the bottle is quite cold before you start. You could put it in the refrigerator for half an hour.

❷ Moisten the edge of the lid thoroughly with water and place the lid on top of the bottle, but don't screw or press it down. The water makes an airtight seal.

❸ Hold the bottle in both hands. Stand still and watch the lid carefully. What happens?

Elastic power

You will need

- **an adult to help**
- plastic bottle with lid
- elastic band
- sticky tape
- paperclips
- bead
- spent matchstick
- pencil

1 Ask an adult to make a hole in the lid and bottom of a plastic bottle.

2 Stretch out a paperclip and push it through the lid, bending it up again at the bottom so it can't slip out.

3 On the other end of the paperclip thread a bead and bend the top over again. Push a pencil through the top as shown.

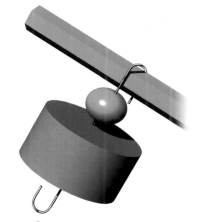

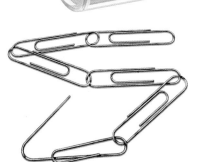

4 Thread an elastic band through the bottom of the bottle and push a matchstick through. Use tape to secure it. Now use lots of paperclips joined together to "fish" the other end of the elastic band and pull it up to fit over the paperclip in the lid.

5 Screw on the lid and wind the pencil to put lots of twists in the elastic band. Then put the bottle down and let it go. What is powering your moving machine?

Hot and hotter

You will need

- paper bag
- hairdrier

1 Blow into a paper bag and throw it up into the air. What happens?

2 Fill the bag with hot air from a hairdrier (don't touch the paper with the drier) and repeat your throw. What happens now?

Electrical charges

Electricity is the movement of tiny particles of matter that are electrically charged. They can have a positive or a negative charge. Just as unlike poles of magnets attract each other, so do unlike electrical charges, while two positive charges, for example, repel (push away) each other. Static electricity is what sometimes gives you a tiny shock when you walk on a nylon carpet.

EXPERIMENT Balloon bends water!

You will need
- balloon
- water from a tap

❶ Turn on a cold tap so that there is a gentle but steady stream of water coming from it.

❷ Rub a balloon on your sweater or some cloth to give it an electrical charge.

❸ Gently move the balloon towards the water and watch what happens.

❹ Let the balloon touch the water and move it away again. Does the effect still happen? What do you think you could do now to make it work again?

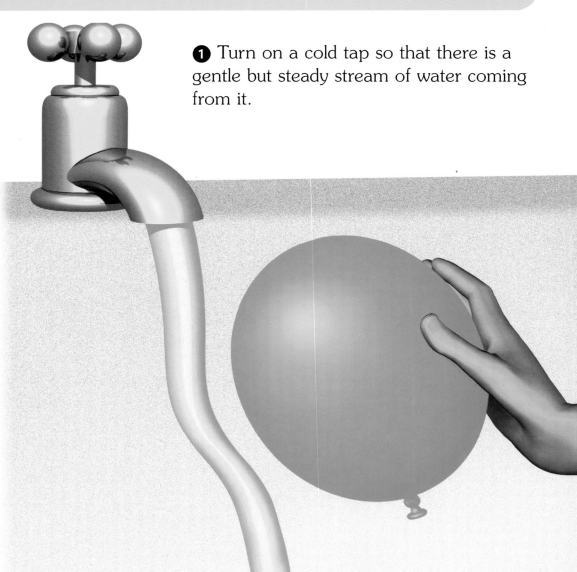

EXPERIMENT Hair-raising!

You will need
- a volunteer with fairly short, clean hair
- balloon

❶ Rub a balloon on some cloth to give it an electrical charge.

❷ Hold the balloon above a friend's head and watch what happens. If you are experimenting on yourself, use a mirror!

EXPERIMENT Charged or not?

You will need
- 2 plastic pens
- sewing thread
- small objects made of wood, glass, plastic and metal
- different kinds of fabric

❶ Tie thread around a plastic pen and hang it up so that it turns freely.

❷ Rub each small object with each of the fabrics in turn and bring it close to the pen. If the object is charged, the pen will swing towards it. Silk, artificial fibres and fur often work well.

EXPERIMENT Attractive balloons

You will need
- 2 balloons
- sewing thread
- sheet of paper

❶ Tie a long piece of thread between two balloons. Rub the balloons to charge them and hold them up by the middle of the thread. Are the balloons attracted to each other?

❷ Put a piece of paper between the balloons. What happens? Objects with the same charge repel each other. Something put close to a charge gains the opposite charge. Opposite charges attract.

Electrical energy

Batteries are devices that store energy. When batteries form part of a circuit, energy in the form of electrons flows out of the negative electrode of the battery, through the circuit and back into the positive electrode. Materials that do not conduct electricity very well are still very useful. They are called resistors and make it possible to dim lights, for example.

EXPERIMENT Make your own battery

You will need
- strong, absorbent kitchen paper
- kitchen foil
- scissors
- copper coins
- salt
- water
- 2 plastic-covered copper wires
- small light bulb

1 Dissolve some salt in water.

2 Cut some pieces of absorbent kitchen paper and kitchen foil so that they are just slightly larger than your copper coins.

3 Dip the pieces of kitchen paper in the salty water.

4 Make a sandwich on the table, starting with a copper coin, then a piece of foil, then a piece of kitchen paper, then a coin, and so on, until you have a good pile.

5 Tuck one exposed end of one wire under the pile and fix its other end to the light bulb. Fix the other end to the top of the pile and the light bulb. What happens?

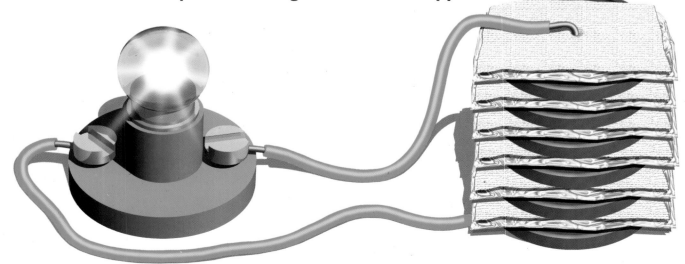

EXPERIMENT Make a switch

You will need
- 4.5V battery
- small light bulb
- 2 wires
- metal paperclip
- 2 drawing pins
- cork mat or board

1 Set up a circuit as shown. One end of the paperclip is around a drawing pin. The other end is free but able to touch the second drawing pin.

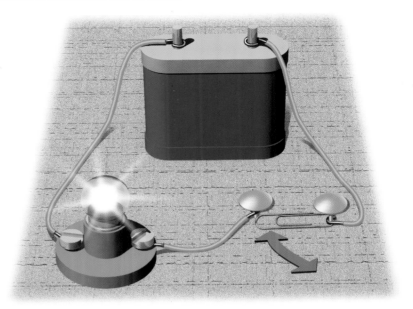

2 The paperclip forms a switch that can complete or break the circuit. Open and close your switch and see what happens.

EXPERIMENT Test for conductivity

You will need
- the items for the experiment above
- small metal, plastic, wood, glass, rubber and cork objects

1 Set up the circuit as above but replace the paperclip with each of the small objects in turn to complete the circuit.
2 Those that allow electricity to flow through them and light the bulb are said to be good conductors. Those that do not are called insulators.

EXPERIMENT Showing resistance

You will need
- **an adult to help**
- circuit as above
- 2 pencils
- pencil sharpener

1 Sharpen both ends of one pencil. Ask an adult to cut or snap the other pencil in half. Sharpen both ends of one half of this.

2 Put each pencil in turn between the drawing pins (moving them as necessary). The pencil lead is made of graphite, which does conduct electricity but not as well as wire. The brightness of the bulb shows if the longer or the shorter pencil conducts electricity better.

A question of gravity

The force of gravity is working on your body all the time. Your centre of gravity is the point from which all the weight of your body seems to act. For stability (keeping your balance), your centre of gravity needs to be over your feet when you are standing. Skiers can lean forward on slopes because their "feet" (the skis) are so long that their centre of gravity is still over them.

EXPERIMENT A downward force

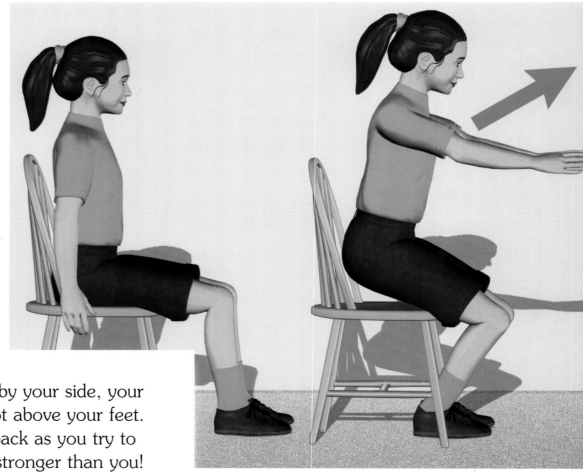

You will need
● chair

❶ Sit upright on a chair with your arms by your sides and position your feet so that they are flat on the floor but slightly in front of your knees, *not* under the chair.
❷ Now, without moving your arms, try to get up.

When your arms are by your side, your centre of gravity is not above your feet. Gravity will pull you back as you try to rise – and it is much stronger than you!

EXPERIMENT Heavy base

You will need

- **an adult to help**
- old tennis ball
- plastic bottle of the same diameter as the tennis ball
- modelling clay
- paints, card etc. for decoration
- sticky tape

❶ Ask an adult to cut an old tennis ball in half and to cut the top off a plastic bottle, about 10cm (4 in) from the top.

❷ Decorate the bottle to look like a person or animal, adding cardboard, paper, woollen hair or anything that is not too heavy.

❸ Fill the tennis ball with modelling clay, pressing it down until it is flat across the top. Then use sticky tape to attach the ball to the bottom of the bottle.

Even with a round bottom, your figure stands up because it is heavy at the bottom and its centre of gravity is very low. When you push the top right over, the centre of gravity is still over the base, so it bounces up again.

EXPERIMENT Balancing on a point

You will need

- plastic bottle with lid
- modelling clay
- nail
- 2 forks

❶ Push a nail into a small ball of modelling clay. Then push two forks into the clay, one on either side.

❷ Now balance the whole thing on the point of the nail on top of a bottle. The weight of the forks means that the centre of gravity is under the point of the nail, so however you push it, it will not overbalance. Tightrope walkers use the same idea when they hold a long pole.

69

Powerful pivots

Machines make work easier by enabling a small force to move a large load. A lever is a simple machine. It pivots around a point called a fulcrum. By changing the position of the fulcrum and the load being moved, greater force can be applied or the load can be moved through a greater distance. Oars are levers that enable a rower to move a huge load of water with every stroke.

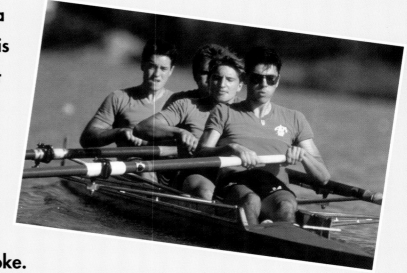

EXPERIMENT Lifting with levers

You will need
- empty yogurt pot
- water
- 30cm (12 in) ruler
- string
- triangular toy brick

❶ Tie some string around the top of the yogurt pot to make a handle or ask an adult to make some holes to thread the string through. Fill the pot with water and try lifting it with just one finger. Is it easy?

❷ Now set up the lever and pivot as shown below. Balance the yogurt pot on one end (make it more stable with some modelling clay if you like) and press down with one finger on the other end. Was this way of lifting the pot easier? Why?

Distance and force

You will need
- 30cm (12 in) ruler
- cereal packet
- 2 triangular toy bricks
- small plastic bag
- marbles

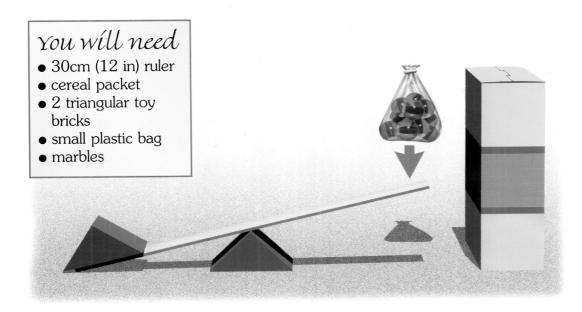

1 Fill a plastic bag with marbles and tie the top. Set up the experiment as shown. Using the top of the cereal packet as a guide, drop the marbles on to the end of the ruler (lever). What happens to the brick on the other end?

2 Now move the triangular brick (the fulcrum) further from the cereal packet. What happens when you drop the bag from the same height now? Experiment with moving the fulcrum to other positions and dropping the bag further from the end of the lever.

Mini-weights

You will need
- drinking straw
- scissors
- dressmaker's pin
- cardboard
- pencil

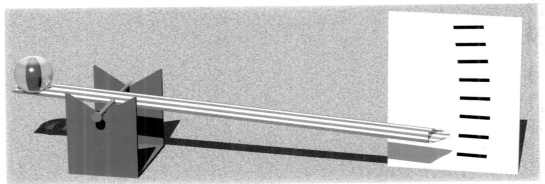

1 You can use the principle of the lever to compare the weights of tiny things such as grains of sugar or seeds. Cut one end of a drinking straw into a scoop shape. Cut the other end to a point.
2 Cut a stand for the balance from card. Push a pin through the straw near the scoop end and balance it on the stand, making sure it can pivot freely. Prop up a piece of white card at the pointed end of the straw. Now try placing tiny objects in the scoop end and marking on the card where the pointed end stops.

71

Pulley power

Pulleys are simple machines that are really wheels with grooves in them through which a rope can run. It is much easier to use your body weight to pull something down than it is to try to push the same thing up. Pulleys make it possible to use a downward force to lift an object up. Most lifts work on a pulley system, with a counterweight that falls as the lift rises.

EXPERIMENT One against two

You will need
- 2 friends
- 2 strong poles or broom handles
- rope or very strong string

❶ Set up the system you see below.
❷ Two of you hold on to the poles and try to pull them apart, while a third pulls on the end of the rope and tries to pull them together. Are two people stronger than one?

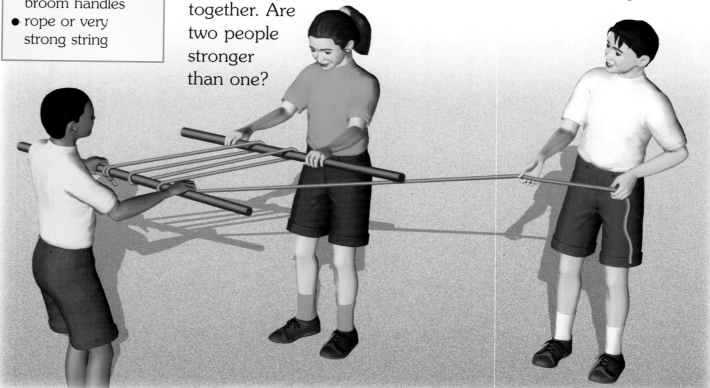

Make a crane

- **an adult to help**
- 2 cotton reels
- wire from a coathanger
- 2 cup hooks
- engineering brick
- thin rope or strong string

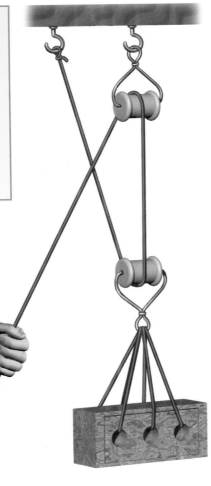

1 Ask an adult to cut two lengths of wire from a coathanger, thread them through the cotton reels, and bend and twist the ends as shown.

2 Ask the adult to screw two hooks into the top of a door frame or other suitable place with a space below. A shed might be a good place.

3 Carefully try to lift a brick (it can be any kind of brick, but engineering bricks have useful holes in them!) from the table with one hand. It is very difficult.

4 Now set up the pulley as shown. Pull down gently on the rope. How hard is it to lift the brick now?

EXPERIMENT # Make a model lift

- **an adult to help**
- 4 cotton reels
- wire from a coathanger
- 2 cup hooks
- 2 bricks
- thin rope or strong string

1 Follow the instructions for the experiment above but this time make a more complicated pulley like the one in the picture.

2 Hang a brick from the bottom of the pulleys and from the rope end. Your bricks are a lift and its counterweight. By gently putting pressure on one brick, you should be able to raise the other brick. The counterweight means that less force is needed to raise the rising brick. In a lift shaft, a counterweight reduces the electrical energy needed to operate the lift.

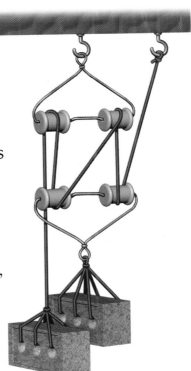

Gearing up

A gear is simply a wheel with teeth. These teeth can fit into the teeth of another gear so that when the first turns, the second does, too. In fact, any number of gear wheels can be fitted together. By using different sizes and numbers of wheels, the direction and the size (magnitude) of a force can be changed. Try these experiments to find out how for yourself.

EXPERIMENT Making gear wheels

You will need
- **adult to help you**
- thick cardboard (such as from a large cardboard box)
- corrugated cardboard
- saucer or lid
- ruler
- pencil
- scissors
- glue
- map pins
- notice board

❶ Draw around a saucer or lid eight times on to thick cardboard. Ask an adult to cut out the shapes for you.

❷ Glue the circles together in fours to make two thick wheels. Measure the thickness of your wheels and draw strips of that width on the back of corrugated cardboard, going *across* the corrugations. Cut these out *very* carefully with scissors.

❸ Glue the strips of corrugated cardboard around your wheels, matching the ends carefully if you need to make joins.

❹ Place the wheels on a noticeboard so that the "teeth" fit together. Push a map pin through the centre of each. Glue a tab on each to act as a handle for turning. If you turn one wheel clockwise, which way does the other wheel turn?

74

EXPERIMENT Bicycle gears

You will need
- **an adult to help**
- everything from the last experiment
- large piece of card
- elastic bands

Gears on a bicycle change the way in which the force of your foot on the pedal turns the back wheel. Make these models to see what happens.

❶ With the help of an adult, make one gear wheel about 20cm (8 in) in diameter, one of 10cm (4 in) and one of 5cm (2 in).

❷ Fix the largest wheel on to a board as shown. It represents the pedal. Glue the second wheel in the centre of a large circular piece of card. It represents the largest gear on the back wheel. Link the two gears with an elastic band, or several elastic bands tied together.

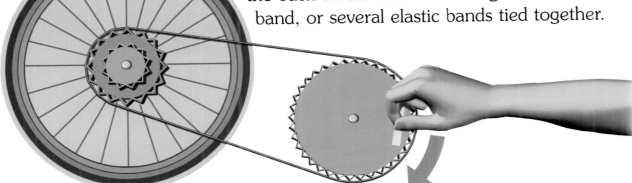

❸ Make a mark at the top of the "bicycle wheel". How many turns of the "pedal" bring the mark to the top again?

❹ Glue the smallest gear wheel on top of the medium-sized one. Reposition the elastic band. You may need to place some card beneath the "pedal" to raise it as well. How many turns of the "pedal" move the wheel one full circle now?

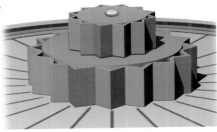

EXPERIMENT Up or down?

You will need
- everything for the experiment on page 74
- paints and brush

❶ Use what you have learnt to make a complicated system of gear wheels. The wheels can be any size or colour you like, as long as they all turn when you move the first one.

❷ Ask friends to tell you whether an arrow on the last wheel will move up or down when you turn the first wheel. They will always get it wrong if they don't think to ask you which way you will be turning!

75

Smart slopes

It may sound strange to say that a slope is a machine, but just like a lever or a pulley it is something that makes work easier. Moving a load from one point to another always takes the same amount of work, but if you spread the work out over a longer distance, less force is needed. Pushing a load up a ramp is easier than lifting it. Screws, axes and spades are kinds of slope, too.

EXPERIMENT Lifting a car!

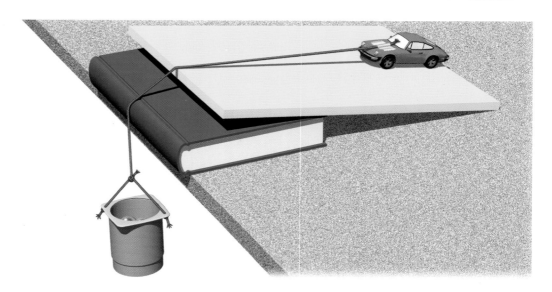

You will need
- chopping board
- books
- small toy car
- string
- empty yogurt pot
- marbles
- sticky tape

1 Use sticky tape to attach one end of a piece of string to the underside of a small toy car. Set up an experiment as shown in the picture. Use just one book to make the slope of your chopping-board quite gradual. Now add marbles to the yogurt pot until the car begins to move up the slope.

2 One by one, add more books to make the slope much steeper. Do you need to use more or fewer marbles to make the car move up the slope?

With just one book, the car was not lifted far. Little force (fewer marbles) was needed. When the car was lifted higher, more force was needed.

EXPERIMENT   How screws work

You will need
- **an adult to help**
- wood screw
- piece of softwood
- screwdriver
- pencil
- string
- sticky tape
- scissors

1 Ask an adult to make a small guide hole for you in a piece of softwood, using an awl or by hammering a nail in a short way and removing it. Can you now push a wood screw all the way into the wood? It would take enormous force to do this.

2 Now try using a screwdriver to turn the screw until it is pushed into the wood. It is much easier to do this.

3 In fact, screws are slopes wrapped around cones. Attach a piece of string to one end of a pencil with sticky tape and wrap the string around it like the thread of a screw. Cut the string when you reach the bottom.

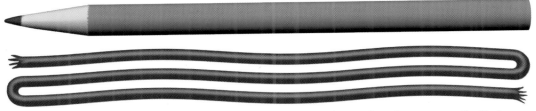

4 Now unwrap the string and lay it out beside the pencil. It is much longer. This is the distance that a screw of the same size would travel. Less force is needed to travel a longer distance.

EXPERIMENT   Wonderful wedges

You will need
- cube toy brick
- triangular toy brick with same size base as the cube
- modelling clay
- heavy books

1 Make a slab of modelling clay. Put the cube and the the triangular brick upside down on top. Balance the same weight on each.

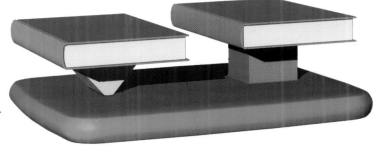

2 Look at the marks the bricks have made. The triangular brick works as a wedge – a kind of slope. Downward pressure forces the surface of the clay open, making a deeper dent. This is how axes work – a downward force splits wood sideways.

Forceful friction

One force that cannot be ignored when you are thinking about movement is friction. This is the force that slows the motion of two surfaces when they move across each other. Engineers spend a lot of time trying to reduce the friction in engines, so that they work more efficiently. But friction can also be a useful force, between the tyres of a car and the road, for example.

EXPERIMENT Slide the slope

You will need
- selection of shoes
- plank of wood or flat, rectangular tray
- small plastic bag of sand or earth to act as a weight
- long ruler
- notebook
- pencil
- friend to help you

We need shoes that grip the ground well. This is particularly important in fast-moving activities such as sports. Before you start the experiment, guess which shoe will have the greatest grip (friction) between its sole and the plank.

❶ Lay a plank on a table or floor. Ask a friend to hold a ruler upright at the end of the plank.

❷ Place a shoe at the ruler end of the plank as shown, with a bag of sand or earth inside it to weigh it down a little. Slowly lift the end of the plank until the shoe begins to slide. Make a note of the type of shoe and the height of the plank's final position. Test the rest of the shoes. Was your guess correct?

EXPERIMENT Feeling friction

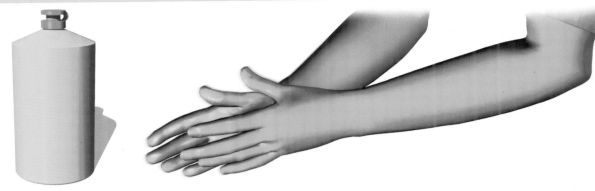

1 You can test the effects of friction very simply by rubbing your hands together for a minute or two. Make sure they are clean and dry first. Friction creates heat, which is why your hands feel warmer.

2 Smoother surfaces create less friction. Put a small amount of washing-up liquid on your palms and rub them together again. Does it take longer for them to feel warm? The washing-up liquid acts as a lubricant. Oil does the same job in engines.

EXPERIMENT Measuring friction

You will need
● wooden board
● 4 map pins
● 4 drawing pins
● sandpaper
● CD case
● thin string
● clean, empty
   yogurt pot
● marbles
● notebook
● 2 pencils

Here is a way of measuring and comparing the friction caused by different surfaces.

1 Arrange a wooden board, map pins and one pencil on the edge of a table as shown. The pencil should be able to turn freely.

2 Knot one end of the string and shut it into the CD case, fixing it with tape if necessary. Tie the other end around the yogurt pot.

3 Put the CD case at the far end of the board. Place marbles in the yogurt pot one by one until the case slides to the pencil. Make a note of the number.

4 Pin sandpaper on to the board and repeat. Which surface created more friction and needed more marbles?

Machines at speed

To move at speed, it is not only friction between moving parts that needs to be kept to a minimum in machines. Friction between the outside of the the moving object and the air itself also has to be reduced as much as possible. An aerodynamic shape, with smooth outlines and as small a surface area as possible, means that higher speeds can be achieved.

EXPERIMENT Spinning in space

You will need
- office chair that can spin around
- friend to help you
- lots of space!

❶ Put the chair in the middle of a clear space. Sit on it and ask a friend to start spinning you around as fast as possible.

❷ Warn your friend to stand clear. What happens when you stretch out your legs and arms, then quickly tuck them in again?

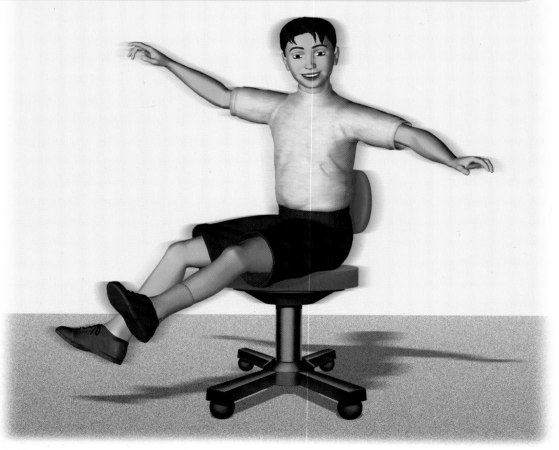

When you push your legs out, air resistance slows you down. You need a smaller, aerodynamic shape to speed up again.

Humming button

- large button
- thin, strong string, thread or twine

1 Thread some string through two holes in a large button and join the ends to make a long loop as shown. Holding the ends loosely, swing the button around and around to wind up the string.

2 Suddenly pull your hands apart to make the button spin. If you keep loosening and tightening the string, the button will move so quickly that it will vibrate the air fast enough to make a humming sound.

EXPERIMENT # Up, up and away

You will need
- large balloon – long, not round
- thin, strong string
- plastic ballpoint pen
- masking tape

1 Blow up a long balloon and tie the end with string.
2 Carefully take the top and middle out of a ballpoint pen to leave just the case. Tape the case to the side of the balloon.

3 Thread a length of string through the pen case and use masking tape to fix the string between the floor and ceiling of a room. Ask an adult first and try to get the string as taut as possible.
4 Hold the balloon near the floor, untie the string around the end, and let the balloon rocket to the ceiling!

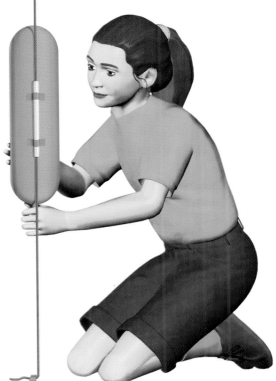

Glossary

Aerodynamics The way in which bodies move through the air. The shape and smoothness of a body effects how air flows over it and can increase or decrease air resistance.

Counterweight A weight that balances another weight. Counterweights are used to reduce the force needed to raise a lift, for example. The force of gravity on a counterweight attached to the lift by a pulley, means that less energy is needed to work the lift.

Electrical charges Forms of negative or positive energy created by the movement of electrons.

Force Something that makes a body move if an equal or greater and opposite force is not present. A force can also cause a body to change shape, speed up or slow down.

Fossil fuel Oil, coal or gas created when the bodies of prehistoric animals were crushed under great pressure millions of years ago. The Earth's supply of fossil fuels is limited.

Friction A force that is caused by the movement of two surfaces against each other. Friction produces heat and tends to slow down the movement of the two surfaces.

Fulcrum The pivot around which a lever turns when a force is applied to it.

Gravity Any large body attracts other bodies towards it. The force exerted by a planet such as Earth is called gravity. It is the even greater gravity of the Sun that keeps the planets of the Solar System in orbit around it.

Inertia The tendency of a body to keep still or to keep moving in the same direction unless a force acts upon it.

Reaction An opposite but equal force caused by another force. All forces cause equal and opposite reactions.

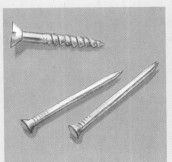

Music
and Sound

Good vibrations

There is no sound in space because there is no air through which vibrations can pass. Unlike light, sound needs to travel through another medium by making its molecules vibrate rhythmically. The vibrations are passed from one molecule to the next. There are lots of ways of making sounds: striking, twanging or blowing is used in most musical instruments.

EXPERIMENT Bouncers

You will need
- plastic bag
- plastic bowl
- large elastic band
- dry salt or sand
- baking pan
- wooden spoon
- friend to help

❶ Put a plastic bowl inside a plastic bag and pull it as taut as you can.
❷ Fix an elastic band around the bowl to hold the plastic in place.
❸ Hold the bowl by the plastic as shown and sprinkle a few grains of salt or sand on top.
❹ Ask a friend to hold a baking pan near the bowl and bang it with a wooden spoon. What happens to the grains on the plastic?

Sound vibrations travel through the air and vibrate the plastic over the bowl – and your ear drums!

Twangers

You will need
- plastic, wooden or metal ruler
- table

1 Hold a ruler over the edge of a table as shown and twang the ruler with your other hand.

2 What happens if you make the part you are twanging longer? What happens if it is shorter? What happens if you twang harder or more gently?

EXPERIMENT Elastic guitar

You will need
- wooden board or pin board
- drawing pins
- elastic bands
- metal paperclips
- sticky tape
- 2 pencils
- yogurt pots
- string
- marbles

1 Tape two pencils to a board about 15cm (6 in) apart. Push four or six drawing pins into the board above the top pencil.

2 Tie string around the yogurt pots and loop it over the top to make a handle. Use paperclips as hooks.

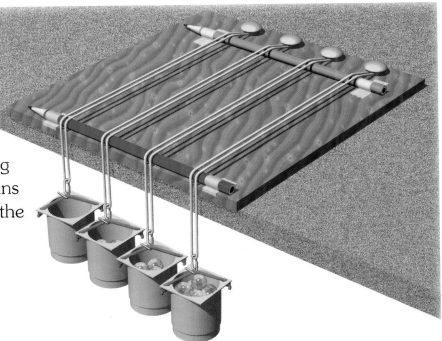

3 Assemble your guitar by hooking elastic bands around the drawing pins and hanging the yogurt pots from the other end of the bands. Put some marbles in each pot.

4 Now try adding and subtracting marbles to tune your guitar.

Sound travels

All sounds travel at a speed of 343m (1,125 ft) per second at room temperature. Light can travel at up to 300,000km (186,000 miles) per second, so from a distance you may see the smoke from the firing of a cannon, for example, before you hear the sound. Sound vibrations do not only travel through air. Solid materials can transmit them more clearly.

EXPERIMENT Distant drums

You will need
- a friend to help
- large metal tin or drum
- strong stick to hit it with

❶ Find a large open space. Give your friend the "drum" and walk away from each other, at least two hundred metres (yards).
❷ Your friend should hit the "drum" as loudly as possible. Walk towards him or her until you can just hear it. (If you can hear it very well, walk a bit farther away until it is fainter.)

❸ Now look carefully as your friend plays again, this time using big, exaggerated movements. Does it seem as if the drum is hit before you hear the sound? If you can borrow a pair of binoculars, you will be able to see even more clearly.

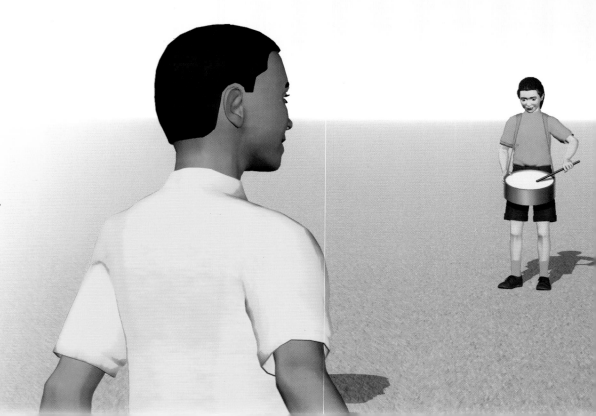

EXPERIMENT Secret messages

You will need
- two friends
- house wall

1 Each friend should invent a different "code" of long and short sounds that can be tapped out on a flat surface. Make sure you know what they are.

2 Now position yourself as shown, around a corner with your ear to the wall. Ask your friends to tap out their signals on the palms of their hands. You will probably not be able to hear them. Now ask them to tap them out on the wall. Can you tell which friend is tapping?

EXPERIMENT Tin-can telephone

You will need
- **an adult to help**
- 2 empty, clean food cans
- nail
- hammer
- strong string

Ask an adult to help you make sure there are no sharp edges on the cans before you use them.

1 Ask an adult to make a hole in the end of each can with a nail and to hammer the edges of the hole so they are flat and safe.

2 Knot a piece of string and thread it through one hole. Then thread it through the other can and knot it again.

3 Test your tin-can telephone. What difference does it make if the string is held taut or slack?

Louder sounds

Sound vibrations don't know they should be heading for your ears! They simply spread out through the air, gradually becoming too weak to hear. But you can direct sound vibrations by making them bounce off solid objects. They cannot escape into the air, so they are still strong when they vibrate against your ear drums. Try these simple experiments to prove this.

EXPERIMENT   Big ears

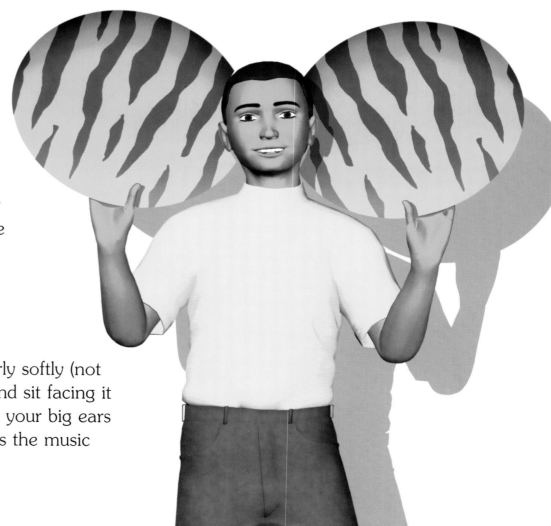

You will need
- cardboard
- scissors
- radio, music centre, CD or cassette player

❶ Could you hear better if you had bigger ears? Cut yourself some big ears out of cardboard, shaped so that you can hold them next to your head behind your own ears.

❷ Play some music fairly softly (not through headphones) and sit facing it as you listen. Now hold your big ears next to your head. Does the music sound louder?

EXPERIMENT Simple stethoscope

You will need
- plastic tube or hose about 1m (3 ft) long
- funnel
- a friend to help

❶ Push a funnel into one end of a 1-metre (3-ft) length of tube or hose. Stand about a metre (3 ft) from your friend. Can you hear his heart beating?

❷ Now ask him to hold the funnel to his chest, while you put the other end of the tube to your ear, still standing the same distance apart. What can you hear now? What has made the difference?

EXPERIMENT Ticking tube

You will need
- watch or clock with a quiet tick
- cardboard tubes of various diameters and lengths
- long ruler

❶ Put the watch on a surface and slowly move your head towards it until you can hear it. Use a ruler to measure how far that is.

❷ Take a cardboard tube longer than the distance you measured and hold it to your ear. Can you already hear the watch?

❸ Experiment with different diameters and lengths of tubes, taping some together to make longer ones. What do you find?

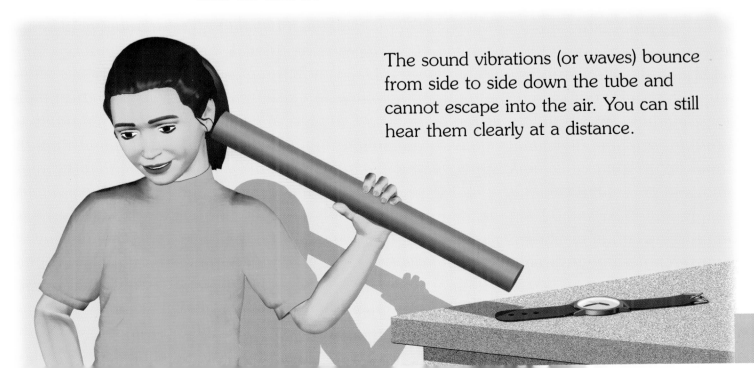

The sound vibrations (or waves) bounce from side to side down the tube and cannot escape into the air. You can still hear them clearly at a distance.

89

Tuning up

The faster sound vibrations follow each other, the higher the pitch of the note they make. If a drum skin is tightened, or stretched over a smaller area, it will vibrate more quickly and make a higher-pitched sound. The closeness of sound vibrations to each other is known as their frequency. This is measured in Hertz (Hz). One vibration per second is one Hertz.

EXPERIMENT   A different drum

You will need
- large plastic bag
- plastic bowl
- elastic band

❶ Put a plastic bowl inside a plastic bag and pull it as taut as you can.
❷ Fix an elastic band around the bowl to hold the plastic in place.
❸ Hold the bowl by the plastic as shown and tap the edge of the bowl to make a sound.
❹ Tighten and loosen the plastic by squeezing and relaxing your lower hand to change the pitch of the note.

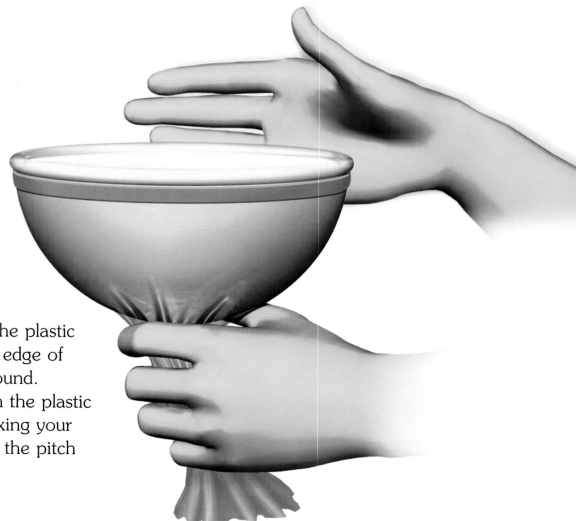

A bottle organ

You will need
- 8 bottles of the same size
- water
- wooden spoon

1 Line up eight bottles and fill each with a little more water than the one before.

2 Strike each bottle with a wooden spoon and listen to the sound it makes.

3 Now try blowing gently across the top of the bottles. The note will be higher, the more water there is in the bottle, because there is less air to vibrate.

4 Adjust the water in the bottles until you have a scale and try playing some tunes!

Pan pipes

1 Cut the end of a straw into a point as shown. Squeeze the cut ends together with your fingers to make them as flat as possible. Put this end in your mouth and blow until you can make a loud, booming note. You may need a little practice!

2 Now make straw pipes of different lengths and join them together to make pan pipes. What happens when you blow two or more notes together?

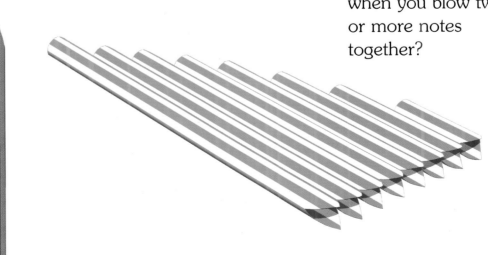

Blowing makes the cut ends of the straw vibrate. This in turn makes the air in the straw vibrate and produce a sound.

Sound signals

Have you ever heard a train whistle as it passes you at speed in a station? The sound drops in pitch as the train passes you. This is called the Doppler effect. It is caused by the fact that when the train is moving towards you, the sound vibrations are bunched up, giving a higher pitch. Sound signals are often used for safety reasons, so understanding how they work is important.

EXPERIMENT | Swooping sirens

You will need
- **an adult to help**
- ruler with a hole in the end
- nylon string or very strong twine

❶ You need to be in a large outside space, well away from other people or buildings. Tie some very strong string really firmly through the hole in the end of a ruler. Ask an adult to check that it cannot come untied.

❷ Twirl the ruler around and around very quickly on about 50cm (20 in) of string. What do you hear? Why isn't the sound the same all the time?

92

EXPERIMENT Intruder alarm!

You will need
- 4.5V battery
- bulb or buzzer
- 2 lengths of plastic-covered copper wire
- cardboard
- scissors
- sticky tape
- kitchen foil

1 Make a circuit like the one below. You could replace the bulb with a buzzer if you like. Tuck the folded card just under a rug so that anyone coming through the door would step on it.

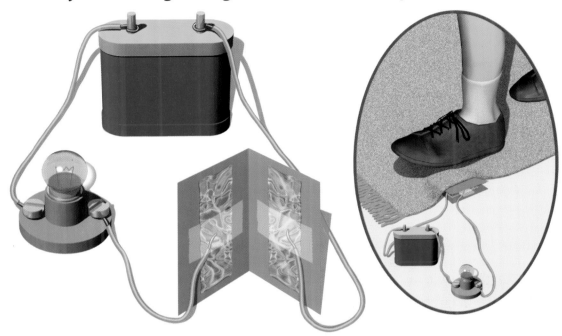

2 When the two pieces of foil touch, they complete the circuit and the bulb lights or the buzzer goes off.

EXPERIMENT Morse code

You will need
- a friend to exchange messages with
- two copies of the Morse code opposite
- paper and pencil

1 There are lots of ways of using the short and long signals of Morse code to send messages. You could tap on a wall or a pipe, bang a drum or even blow a whistle. In fact, you don't need to make a sound at all! At night, Morse messages can be sent by flashing a torch. Try sending short messages to a friend to start with.

2 Write the signals down as they arrive and work out what they say later. After a while, you may be able to interpret the signals as they arrive.

A	• — —	S	• • •
B	— • • •	T	—
C	— • — •	U	• • —
D	— • •	V	• • • —
E	•	W	• — —
F	• • — •	X	— • • —
G	— — •	Y	— • — —
H	• • • •	Z	— — • •
I	• •		
J	• — — —	1	• — — — —
K	— • —	2	• • — — —
L	• — • •	3	• • • — —
M	— —	4	• • • • —
N	— •	5	• • • • •
O	— — —	6	— • • • •
P	• — — •	7	— — • • •
Q	— — • —	8	— — — • •
R	• — •	9	— — — — •

Sound and silence

Different materials have different effects on the sound vibrations that hit them. They may reflect the sound waves, or they may absorb them. We say that they have different acoustic properties. Architects designing concert halls have to use a variety of materials to make sure that music is neither lost in the walls and ceiling nor echoed around the room.

EXPERIMENT Resonators rule

You will need
- plastic food box
- hardback books of a similar size
- 2 pencils
- 2 large elastic bands

❶ Take a book that is roughly the same size as a plastic food box. You can put two books on top of each other if your box is deep. Place a pencil at either end of the book and stretch an elastic band over them as shown.

❷ Stretch an elastic band over the open box, too. Twang both bands, which are stretched very similar amounts. What do you find?

The vibration of the band over the books does not make the solid books vibrate in sympathy (resonate), but the band on the open box is able to vibrate the box and the air inside it, making a fuller sound. Guitars, violins and other stringed instruments have strings mounted on hollow "boxes" for just this reason.

EXPERIMENT Muffling sounds

You will need
- alarm clock or kitchen timer
- cardboard shoebox
- different materials: plastic bags, bath sponges, dried lentils, newspaper etc.
- large room
- watch

❶ Set the alarm or timer to go off in one minute and place it in a shoebox with the lid on at one end of the room. Walk to the other end of the room and, when the minute is up, walk towards the box until you can hear it (you may not need to walk at all!)

❷ Now try to muffle the sound as much as you can, using different materials to do so. The nearer you are to the box when you hear the sound, the better the muffling. You should find that the most successful materials are those that soak up the vibrations, such as bath sponges, rather than those that will vibrate vigorously themselves.

EXPERIMENT Talk back

You will need
- a friend to help
- 2 umbrellas

❶ In an open place outside, stand 2m (6ft) from a friend and try to talk to each other without raising your voices, just as you would indoors.

❷ Now hold an open umbrella behind each of your heads and try again. Your voices will be bounced off the umbrellas so that you can hear each other better.

Hearing and listening

Human ears need good conditions for the brain to make sense of sounds properly. Standing close to someone you are talking to, facing them, with no other distracting sounds nearby, gives the best chance of words being heard correctly. But if a car zooms past, a friend starts to talk at the same time, or you are too far away, understanding becomes much more difficult.

EXPERIMENT A pin drop

❶ We sometimes say somewhere is quiet enough to hear a pin drop. How quiet is that? Stand near a table with your back to it and ask a friend to drop a pin on to it. Can you hear it?

❷ Keep moving one step away from the table and see how far away you have to be before you can no longer hear the pin drop. Your friend might like to pretend to drop the pin sometimes to check your claim!

❸ Now try the same experiment with a radio playing softly on the table. How much more difficult is it to hear the pin now?

You will need
- friends to help
- pencil
- paper
- large room

❶ Sit in front of one wall of a large room, facing the wall. Make a rough plan of the room and your position in it on a piece of paper. Ask your friends to stand behind you, spread out over the room, but don't look to see where they are.

❷ One by one, your friends should clap their hands once loudly. Mark on your piece of paper where you think the sound was coming from. When everyone has clapped, check how accurate you were.

❸ Face front again and ask your friends to move to new positions – quietly! Now put the hand you don't write with over one ear and try the experiment again. How successful are you this time?

EXPERIMENT Megaphone

You will need
- a friend to help
- cardboard
- sticky tape
- scissors

❶ Fold some cardboard into a large cone shape and secure it with sticky tape, trimming the edges with scissors. Snip off the pointed end to make a hole about 5cm (2 in) across and make a cardboard handle.

❷ In a large open space, move away from a friend until you can only just hear each other shouting. Now try using the megaphone. Is it easier to get your message across? The megaphone vibrates with your voice and increases the force of the vibrations reaching your friend.

Believing your ears

Many animals have much more acute hearing than we do. Dogs, for example, can hear sounds up to 50,000Hz, while human beings can only hear up to 20,000Hz. Our ears can often mislead us. We may also not be aware of how much we use our other senses, especially sight, to give meaning to the information coming into our ears. Try these experiments to hear for yourself!

EXPERIMENT Tape test

You will need
- lots of friends
- tape recorder
- favourite rhyme

❶ Choose a short rhyme and ask lots of your friends in turn to repeat it. Record them with a tape recorder. Make sure you record yourself somewhere in the middle of the test.

❷ Listen to the tape. Is it easy to recognize everyone who speaks, including yourself?

One reason why your own voice sounds different to you is that you usually hear it through the vibrations of your skull and the soft parts of your head. On the tape, you are hearing it through vibrations in the air, giving a different sound. Of course, you are used to hearing everyone else's speech through the air, so they sound almost the same as usual.

You will need
- coconut shells
- gravel or sand
- tray
- shoe
- sheets of paper
- bubblewrap
- metal baking sheet

Here are some simple sound effects you could use for a school play or taped story. Try recording them on a tape machine and playing them back to your friends. Can they guess what they are?

 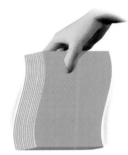

4 A single sheet of paper shaken sounds like just that. But a sheaf of papers held together and shaken gives you the sound of a flag on a flagpole or a ship's sail flapping in the wind.

1 This is an old favourite. Bang half coconut shells together to make the sound of a horse's hooves. Can you make it sound as if the horse is walking, trotting, cantering and galloping?

2 Fill a tray with gravel or sand and "walk" a shoe on your hand across it to make the sound of footsteps. Again, can you make them move at different speeds?

5 Holding a hardback book open and then shutting it very quickly gives a convincing pistol shot. Try different books to get the best sound.

3 Wobble a metal baking sheet to make the noise of thunder. The sheet needs to be the flat kind without sides.

6 Squeezing bubblewrap from packaging rhythmically in your hands sounds like someone walking through snow. Bursting the bubbles at random can resemble twigs crackling on a fire.

Bouncing sound

We sometimes think that waves of the sea push water forward towards the shore. In fact, the water only moves up and down, but the motion passes from one molecule of water to the next. Sound waves are similar. Air does not rush from a musical instrument to our ears, but molecules of air bounce up and down to carry the vibrations towards us.

EXPERIMENT Sound waves

You will need
- baking tray
- water
- dried pea
- large pebble

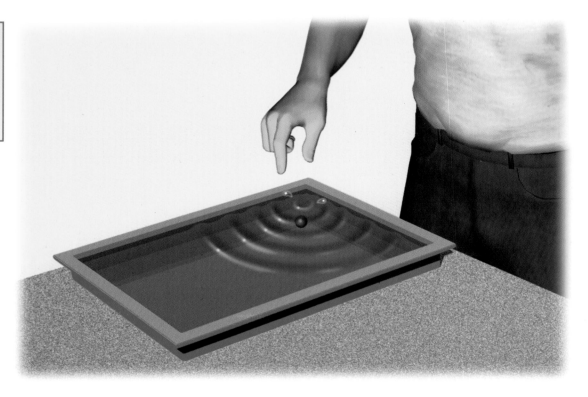

❶ When we make sounds, vibrations pass through the air and other materials to meet our ears. Fill a baking tray with water, place it on a flat surface and wait until the water is perfectly still.

❷ Drop a dried pea into one end of the tray and watch the ripples. These are like the vibrations of a soft sound.

❸ Wait until the ripples are still, then drop a large pebble into the water from the same height. What happens now? These are like the vibrations of a louder sound.

Bouncing back

Why does your singing sound louder in the bathroom, where there are lots of hard surfaces? Try this experiment.

❶ Throw a ball at the base of a wall and catch it as it bounces back. This is what happens when sound hits a solid surface. You may even hear the returning sound as an echo.

❷ Now place a pillow or piece of foam rubber at the base of the wall. Throw the ball at it. The soft material absorbs the energy of the ball, just as soft materials absorb sound energy.

Echo-timing

❶ If you ever find yourself somewhere that sends back an echo of your voice when you shout, you may be able to work out the approximate speed of sound. First measure the distance you are from the foot of the cliff or the back of the cave in metres (or yards).

❷ Now shout loudly and measure on a stopwatch how long it is before you hear the echo. Then double the distance to the cliff or wall and divide it by the number of seconds. That should give you the speed of sound in metres (or yards) per second. The actual speed is 340 metres per second (372 yards per second). How close did you get?

Glossary

Acoustics The study of sound and how it moves around a room. Concert halls and other large public spaces need to have good acoustics.

Doppler effect Named after the Austrian physicist who described it, the rising or lowering of the frequency of a sound as its source and its hearer move towards or away from each other. This is why a train siren seems to change pitch as the train passes.

Echo A sound heard again when it bounces off (is reflected by) a solid object.

Frequency The number of sound vibrations in a second. The higher the number, the higher the pitch of the sound they produce.

Megaphone A funnel-shaped object that amplifies (increases the volume of) a sound passing through it.

Pitch The highness or lowness of a sound, caused by the frequency of the vibrations producing it.

Resonance The increase in length or volume of a sound that happens when it causes other strings, membranes or air inside a box to vibrate in sympathy.

Signal A sign, movement or event that conveys information, such as the sound of Morse Code, the flashing of a light, or the waving of a flag.

Sound waves The way in which air molecules stretch out and bunch together to carry sound vibrations.

Stethoscope An instrument used for listening to the sounds of the heart and other organs. The sound vibrations pass to the ears through tubes and, as they cannot escape, are as loud as possible.

Tuning Adjusting the strings, membrane or other parts of a musical instrument to produce the required notes.

Vibration A rapid backwards and forwards movement of an object or the air. Vibrations can be so small that we cannot feel them – but we may be able to hear them as sounds.

The Air Around Us

Air everywhere

Air is all around us. We could not exist without it. We are so used to breathing it, moving through it and feeling it on our skin that we hardly think about it being there. But air is a gas, or rather a mixture of gases. It has properties just as other materials do and it affects everything we do on Earth. These pages will help you to prove to yourself how amazing air really is.

EXPERIMENT   Is air there?

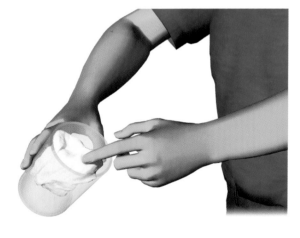

❶ Crumple a tissue and push it into the bottom of a glass.
❷ Fill a washing-up bowl with water.
❸ Turn the glass upside down and push it down into the water until it is covered.
❹ Pull the glass out again, keeping it upside down. The tissue remains dry because there was air in the glass that the water could not push out of the way to reach the tissue.

EXPERIMENT Is air heavy?

You will need
- 2 identical balloons
- rail or pole
- wire coathanger
- 2 plastic pegs
- pin

❶ Blow up two balloons until they are exactly the same size. Tie the ends.

❷ Hang a coathanger up on a rail or pole. You could balance it between two chairs.

❸ Use pegs to attach the balloons to either end of the coathanger. They should balance.

❹ Pop one balloon with a pin. The balloon that is still full of air will cause the coathanger to dip at that side, showing that the air in the other balloon did have a weight that was keeping the coathanger balanced.

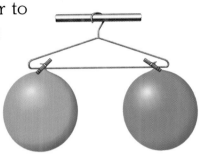 

EXPERIMENT The gases in air

You will need
- **an adult to help**
- glass jam-jar
- candle
- water
- coins
- large glass bowl

❶ Ask an adult to light the candle and drip a little wax into the bowl to fix the candle in place.

❷ Carefully fill the bowl with water.

❸ Place the jam-jar

over the candle. Place piles of coins under it so that its edges come just under the water.

❹ When the candle has burned up all the oxygen in the air in the jam-jar, it will go out. The water will move up into the jar to take the place of the oxygen, showing that about one fifth (20%) of air is oxygen.

Air pressure

We are all under pressure – and not just from parents and homework! The air is pressing down on us all the time, but we don't notice this because we are used to it. Air pressure is constantly changing, bringing about variations in the weather as different areas of pressure make air move, causing winds and storms. A barometer is used to measure air pressure.

EXPERIMENT Paper power

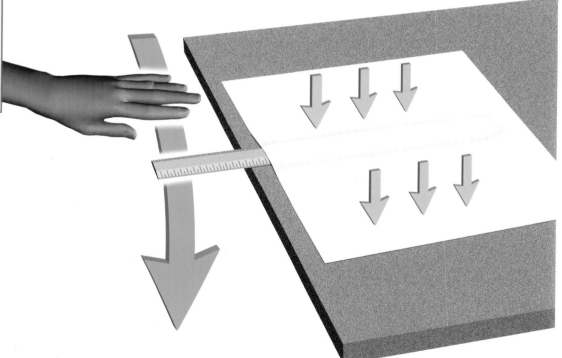

You will need
- table
- ruler
- sheet of newspaper

❶ Position the ruler on the table so that a third of it extends beyond the edge of the table.

❷ Place the newspaper over the ruler, covering as much of it as possible. Smooth out the paper so that there is no air trapped underneath.

❸ Now try to move the ruler by hitting the end with your hand (but don't break the ruler!) The ruler is really hard to move because of the air pressing down on the paper. The large area of the paper means that a lot of air is pressing down on it.

❶ Half-fill a bowl or deep tray and a plastic bottle with water.
❷ Draw a scale on to a strip of paper and use sticky tape to fix it to the side of the bottle.
❸ Position two or three small piles of coins in the bottom of the bowl or tray so that the neck of the bottle can rest

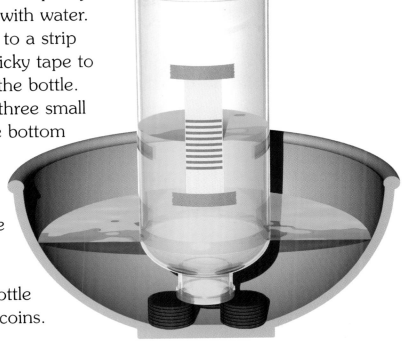

on them. This keeps the neck of the bottle off the bottom of the container so that water can flow in and out.
❹ Put your thumb over the neck of the bottle and carefully stand it upside down on the coins.

Your water barometer shows how air pressure is rising and falling. When pressure goes up, the water will rise in the bottle. When it falls, the water level will fall.

EXPERIMENT Make an air barometer

❶ Cut a balloon open and fix it tightly over the top of a wide-necked jar with an elastic band.
❷ Trim the end of a straw into a point and use tape to fix the other end of it to the centre of the top of the jar.

Rising air pressure = fair weather. Falling pressure = unsettled weather and perhaps rain.

❸ Draw a scale on a piece of card and prop it up at the end of the straw. The air inside the jar expands or contracts as the air pressure rises and falls, making the straw pointer move.

Air temperature

As air heats up, it expands. This means that it becomes less dense than cooler air around it, so the warm air will tend to rise. When air expands in a confined space, it puts pressure on the surfaces around it. The warmth of the Sun causes changes in air temperature and pressure. It is this moving air that causes most of our weather. Try these experiments to find out more.

EXPERIMENT Make a thermometer

You will need

- **an adult to help**
- clear bottle with plastic screw-top lid
- clear drinking straw
- modelling clay
- water
- ink
- hairdrier

① Fill a clear bottle about halfway with water into which you have mixed a few drops of ink.

② Ask an adult to make a hole in the lid of the bottle, large enough for a straw to fit through.

③ Screw the lid on to the bottle and push the straw through. Press modelling clay around the part where the straw goes into the lid to make an airtight seal.

④ Carefully blow into the straw until the liquid comes about halfway up it. Now blow hot air from a hairdrier on to the bottle. What happens to the coloured liquid in the straw? What happens if you put the bottle in the fridge for a while?

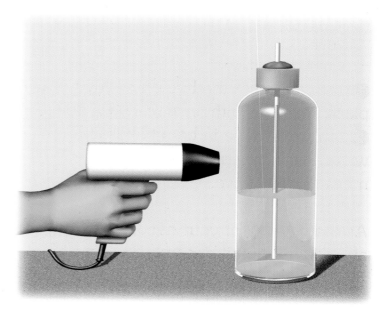

EXPERIMENT Expanding air

1 Blow up three balloons so that they are exactly the same size as each other. Check by comparing the lengths of pieces of string held around the widest part. Number the balloons boldly using a felt-tip pen.

2 Put the first balloon in a hot place, such as an airing cupboard or a greenhouse on a sunny day. Make sure it is not actually touching anything very hot, or it will burst. Put the second balloon in a fridge or freezer. Keep the third balloon at room temperature.

3 After one hour, compare the balloons. What has happened – and why?

EXPERIMENT The heat of your hands

1 Wet the lid of the bottle and simply place it on top without screwing it down.

2 Hold the bottle in both hands and watch the lid. It will suddenly begin to move up and down.

As the warmth of your hands heats the air in the bottle, it expands, pushing off the lid!

Super suction

Suction relies on differences in air pressure. If you drink juice from a cardboard carton, you will see the sides of the carton start to cave in as you finish the drink. You have reduced the air pressure inside the carton. The greater air pressure outside pushes against the walls and makes them collapse inwards. Try these experiments to see some of the surprising results of suction.

EXPERIMENT Silly straws

You will need
- drinking straws
- drinks
- pin

❶ This is a trick you can play on your friends. Before offering them a drink with a straw, use a pin to make a small hole a short distance below the top of each straw. If you do it on the underside, especially if it is in the bendy part of a straw, the hole won't show.

❷ Your friends will suck in vain! Normally, sucking would make the air pressure in the straw less than that pushing down on the surface of the liquid, so the drink would rise in the straw. Because of the hole, the air pressure in the straw does not drop.

You will need

- **an adult to help**
- hard-boiled egg
- glass bottle
- piece of paper

❶ You need a bottle with an opening that is a little bit smaller than the widest part of the egg. First peel the egg! Then screw up some paper and drop it into the bottle.

❷ Ask a grown-up to light the paper with a taper and quickly place the egg, small end down, in the neck of the bottle.

❸ As the paper burns, it uses up oxygen, making the air pressure outside higher than that in the bottle. Watch what happens!

EXPERIMENT Push or pull?

You will need

- drinking straw
- 2 apples
- string
- broom handle
- 2 chairs

❶ Place the chairs about 1 metre (1 yard) apart and lay the broom handle across the seats.

❷ Tie a piece of string to the stalk of each apple and hang them from the broom handle so that they are at the same height and about 1 cm (¹/₂ in) apart. Make sure they are still.

❸ Kneel in front of the apples and use a straw to blow a stream of air between them. What do you think will happen?

Instead of blowing the apples apart, the air seems to suck them together. The fast-moving blown air reduces the air pressure between the apples. As the outer pressure is greater, it pushes them together.

Water in the air

If you climb down into a misty valley or go out on a foggy day, you can almost see tiny droplets of water in the air. But the air contains some water vapour even on the brightest day. Water vapour is invisible until it condenses, which means that it cools and turns from a gas into tiny droplets of liquid. These are what form the steam you see rising from hot water.

EXPERIMENT Make a cloud!

You will need
- clear glass bottle
- hand-hot water
- ice cube
- dark blue or black paper

❶ Carefully fill a glass bottle with hand-hot water.

❷ After three minutes, pour the water away, leaving just a little in the bottom of the bottle.

❸ Place an ice cube on top of the open bottle.

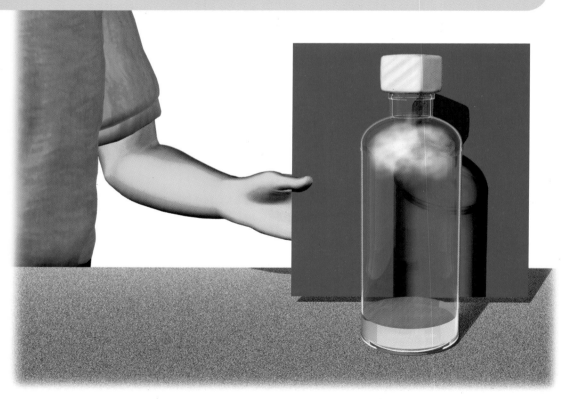

❹ Hold dark paper behind the bottle and watch as a cloud begins to appear where the warm air rising from the bottom meets the cold air at the top. Water vapour in the air condenses to form a cloud of tiny droplets.

EXPERIMENT   It's raining!

You will need
- **an adult to help**
- freezer
- electric kettle
- water
- metal tablespoon
- saucer
- oven mitt

❶ Put a metal spoon in the freezer for half an hour.

❷ Ask an adult to help you with the rest of the experiment. Boil a kettle of water.

❸ Place a saucer under the kettle spout.

❹ Put on an oven mitt and carefully hold the cold spoon in the vapour from the spout (the clear part before the vapour turns to steam). The vapour condenses into liquid when it hits the cold spoon and falls as "rain" to the saucer below.

EXPERIMENT   Make a hygrometer

You will need
- 2 identical thermometers
- cotton wool
- elastic bands
- empty yogurt pot
- water
- large cardboard box without lid
- knitting needle

❶ Use a knitting needle to make two holes 10cm (4 in) apart in a box.

❷ Fix the same amount of cotton wool around two thermometers with elastic bands.

❸ Loop an elastic band around the top of each thermometer and push the other end of the band through one of the holes in the top of the box. Slide the knitting needle through as shown so the thermometers hang freely.

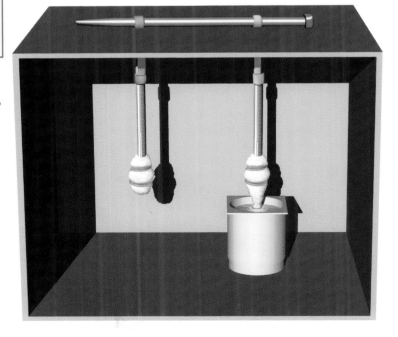

❹ Put a yogurt pot of water under one thermometer, so the cotton wool (but not the thermometer) is always wet.

❺ Compare the readings of the two thermometers at different times of day. The greater the difference in temperature, the less humid it is.

Our windy world

Changes in air pressure caused by the heat of the Sun mean that the air around our Earth is on the move all the time. These experiments will help you to measure the direction and speed of the wind. They need to be done in an open space in a garden. If that is not possible, you may be able to get permission to do them at school in the playground or on a playing field.

EXPERIMENT Make a wind vane

You will need
- **an adult to help**
- long nail
- wooden post (perhaps part of a fence)
- wooden beads
- balsa wood
- hammer
- ruler
- craft knife
- balsa glue
- compass

❶ Ask an adult to cut out pieces of balsa wood as shown below. The slots should be as wide as the thickness of the wood.

❷ Assemble the wind vane as shown, using glue to fix the pieces of balsa firmly.

❸ Balance the vane on the head of a nail to find the centre point. Ask an adult to push the nail through the centre point, threading beads before and after it as shown, and to hammer the nail into a post or fence, making sure the vane turns freely.

❸ Use your compass to find out which way the wind is blowing. The vane points towards the direction the wind is coming from. If it is blowing from the south, it is called a south wind.

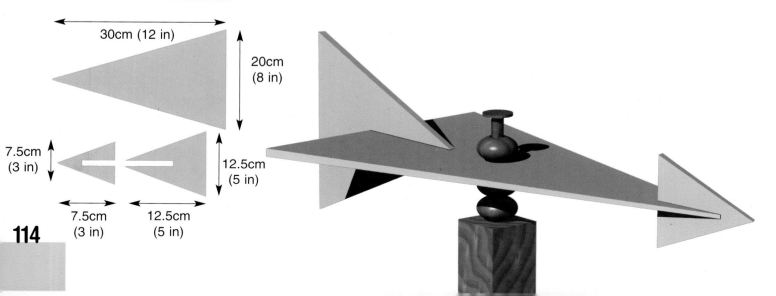

30cm (12 in)

20cm (8 in)

7.5cm (3 in)

12.5cm (5 in)

7.5cm (3 in)

12.5cm (5 in)

A cup anemometer

You will need

- **an adult to help**
- 2 strips of balsa wood 35cm (14 in) long and 1.25 cm (½ in) wide
- long nail
- wooden beads
- 3 white plastic cups or yogurt cartons
- 1 coloured plastic cup or yogurt pot
- ruler
- balsa glue
- post or fence to nail the anemometer to
- hammer
- watch

❶ An anemometer measures wind speed. Find the centre of the two balsa strips and glue them together. Ask an adult to make a hole through them both with a nail.

❷ Glue three white pots and a coloured one to the strips as shown, making sure they all face the same way.

❸ Ask an adult to nail the anemometer to a post in the same way as the wind vane.

❹ To measure the wind speed, you need to count how many times the coloured cup goes past you in one minute.

A spoon anemometer

You will need

- **an adult to help**
- teaspoon
- screwdriver
- wire
- large screw
- sheet of plywood about 20cm x 25cm (8 in x 10 in)
- indelible pen
- ruler
- nails or screws for fixing

The higher the spoon swings, the faster the wind speed.

❶ Turn the screw into the top lefthand corner of the plywood, about 2.5cm (1 in) from the sides.

❷ Wind wire around a spoon handle and the screw, so that the spoon can swing freely and is facing to the left.

❸ Use a ruler to draw a scale on the plywood, then ask an adult to set up the anemometer on a fence or post.

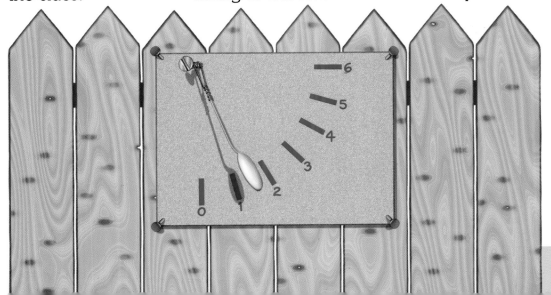

Air in action

The movement of the wind can be used for fun in many sports, such as yachting, sailboarding and even kite-flying. But certain winds can be very destructive. Tornadoes are twisting winds that can reach 500 kilometres per hour (310 miles per hour). They are formed when quickly rising warm air is spun by strong winds. They can suck up everything in their path.

EXPERIMENT ## A tornado in a bottle

You will need
- revolving cakestand
- masking tape
- glass
- fizzy water
- table salt

❶ Tape a glass firmly to the centre of a cakestand.
❷ Fill the glass three-quarters full with fizzy water.
❸ Spin the stand and pour a little bit of salt into the glass.

As the salt sinks into the water, little bubbles of gas are released. The spinning of the glass causes them to form a tiny tornado!

Prevailing winds

You will need
- paper
- coloured pens
- wind vane (see page 114)
- compass

The wind can blow from any direction, but in most places one direction is most common. This is called the prevailing wind. You can discover the prevailing wind in your area by using a wind rose.

1 Draw a "rose" like the one on the right.

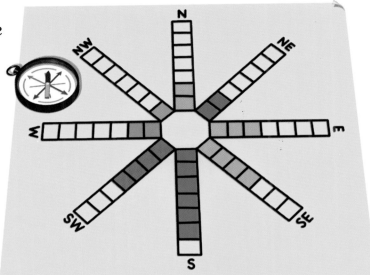

2 Each day, at about the same time, use your wind vane and a compass to check the direction of the wind. Colour in the appropriate square on the wind rose, using a different colour for each direction. If you prefer, you could use coloured stickers instead. After four weeks, check the wind rose. The longest line shows the direction of the prevailing wind.

Wind wanderer

Please let me know where you find this balloon!

You will need
- helium-filled balloons
- string
- addressed labels

Check with your parents that they are happy for you to write your name and phone number on the labels. They may prefer you to put their names or ask if you can use a school address.

1 Launching labelled balloons is a good way of finding out just how far the wind can carry something. You could have a competition with your friends. Simply write labels like the one above and put contact details on the other side. Then launch your balloons and wait!

2 If you get many replies, you could chart where they come from on a map.

Up, up and away

The wing of an aeroplane is of a special shape, called an aerofoil. It is curved, so that the top edge has a larger surface than the underneath. When the wing moves through the air, air above it has further to travel than the air beneath. This means that the air pressure below the wing is greater than that above, causing a force called lift that pushes the wing upwards.

EXPERIMENT Lift in action

You will need
- hardback book
- thin strip of paper

❶ Tuck a strip of paper into a book so that most of it sticks out.

❷ Hold the book with the paper away from you and blow hard over the top of it.

❸ The air moving quickly over the top of the paper will be at a lower pressure than the air below, and the paper should lift up into the air.

Make a model aerofoil

You will need

- **an adult to help**
- paper about 5cm x 25cm (2 in x 10 in)
- sticky tape
- scissors
- drinking straw about 10cm (4 in) long
- strong, straight wire about 30cm (12 in) long
- block of wood
- glue
- hairdrier

1 Ask an adult to make a small hole in a block of wood. Glue a piece of strong wire upright into it.

2 Fold a strip of paper in half. Bend one side into a curved shape and stick it down with tape.

3 Cut a small hole or slot in each side of the paper aerofoil and push a straw through them. Use glue or tape to fix the straw in position.

4 Slide the straw over the wire.

5 Blow air from the hairdrier over the top surface of the aerofoil. When the angle is right, the aerofoil should lift up the wire.

EXPERIMENT # Make a glider

You will need

- strong paper
- sticky tape
- paperclip

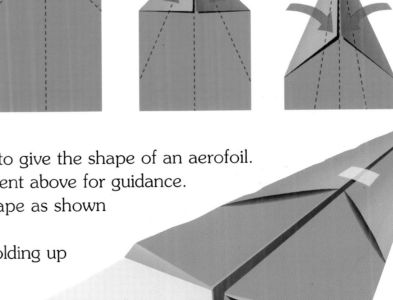

1 Fold a piece of paper, following the steps opposite.

2 Try to curve the wings slightly, to give the shape of an aerofoil. Look at the picture in the experiment above for guidance.

3 Hold the wings together with tape as shown if necessary.

4 Try different wing shapes and folding up the back surface to form ailerons, like the wing flaps of planes. Does positioning a paperclip on the glider improve its flight at all?

119

In the air

For most life to survive, there must be some water vapour in the air. Plants give off water vapour from their leaves. Water that they draw up from the ground through their roots evaporates from the leaf surface. Water in oceans, lakes and rivers evaporates, too. This water forms a cycle – vapour in the air condenses back into clouds, rain and snow when it cools.

EXPERIMENT Plant factories

You will need
- clear plastic bag
- elastic band or freezer-bag tie
- green pot-plant

1 Pull a plastic bag over a vigorous branch of a pot-plant and secure it tightly around the branch with an elastic band or freezer tie.

2 The next day, check the bag. What do you see? Where has the water come from?

EXPERIMENT Evaporation

You will need
- jug of water
- flat, dry surface outdoors
- chalk
- watch

1 Take a jug of water outside on a dry day and pour it on a flat, dry surface such as concrete or tarmac to make a puddle. Choose somewhere where the puddle will not be disturbed by people walking through it.

2 Check the time and draw carefully around the edge of the puddle with chalk.

3 After one hour, repeat step **2**, and continue until the puddle has gone.

4 Where has the water gone? Try the experiment on different surfaces and in different weather conditions. What happens?

EXPERIMENT Condensation

You will need
- **an adult to help**
- window
- drinking glass
- ice
- cold water
- hot water
- bowl

1 Stand close to a window and breathe on it. What appears on the glass? Where did it come from?

2 Pour cold water into a glass and add some ice cubes. Stand the glass in a warm place. What do you see on the outside of the glass. Why?

3 Ask a grown-up to pour some water from a boiling kettle into a bowl. Don't touch it but watch what happens above the bowl. Why does it stop in the end?

Air force

Air pressure is a powerful force. A hovercraft draws in air and forces it out underneath, where it pushes against the ground or sea and reduces friction. This means that less energy is needed to drive the craft forwards. A flexible band of material around the bottom of the hovercraft stops the air from escaping as soon as it is pumped out.

EXPERIMENT Make a hovercraft

You will need
- balloon
- stiff card
- cardboard tube from kitchen roll
- PVA glue
- scissors
- pair of compasses
- string

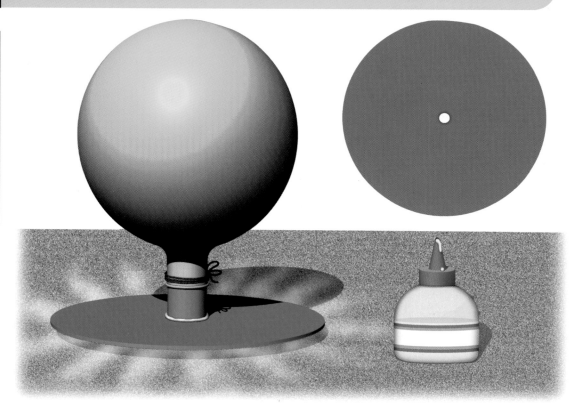

❶ Carefully cut a piece of tube about 5cm (2 in) long.
❷ Use compasses to draw a circle 10cm (4 in) across on stiff card. Cut it out. Make a small hole through the middle of the card with the compasses.

❸ Blow up a balloon and tie it with string about 2.5cm (1 in) from the end. Push this end over the tube, then glue the tube to the card as shown. When the glue is dry, put the hovercraft on a flat surface and release the string.

The weight of air

You will need
- measuring tape
- pencil
- paper
- calculator (if you need one!)

❶ How much does the air in your room weigh? First measure the length, width and height of the room in metres (yards).

❷ Multiply the length by the width. Then multiply the result by the height of the room. Your answer is the number of cubic metres (yards) of air in the room.

❸ Now multiply this result by 1.2 (3.5), which is the approximate weight of one cubic metre (yard) of air in kilograms (pounds). The result is the weight of air in the room. How does it compare with your weight?

❹ As a matter of fact, the air may weigh more or less on different days. Can you think why?

Lung power

You will need
- **an adult to help**
- large empty plastic bottle (such as a water bottle)
- washing-up bowl
- water
- plastic tube
- measuring jug

❶ How much air goes in and out of your lungs when you breathe? You will need an adult to help you find out. Fill the washing-up bowl and the bottle with water. Ask an adult to hold the bottle upside down under the water.

❷ Push a plastic tube or hose up into the bottle.

❸ Take a deep breath and blow into the tube as hard as you can. Bubbles will rise to the top of the bottle. Pinch the tube at once.

❹ Pull the tube out and ask the adult to hold her hand over the neck of the bottle as she turns it upright. To find out how much gas you breathed out, fill the bottle to the top again from a measuring jug and add up how much water you poured in.

Air in use

Not content with breathing it in every minute of the day and night, human beings have found other uses for air. It is vital in cooking and cleaning. By pumping air in and out of special chambers, submarines are able to sink to the sea bed and rise to the surface again. Compressed air is used to power machines, such as those used for digging up roads.

EXPERIMENT Trapped air

You will need
- **an adult to help**
- baking tray
- whisk
- bowl
- spoon
- kitchen scales
- 2 eggs
- 100g (4oz) sugar
- oven
- baking parchment

❶ Ask an adult to help you separate the whites of two eggs from their yolks. Put the whites in a bowl with half the sugar and whisk them vigorously until they stand up in stiff peaks. You have whisked thousands of tiny air bubbles into the mixture. Stir in the rest of the sugar carefully.

❷ Place a piece of baking parchment on a tray and arrange spoonfuls of the mixture so that they are not touching.

❸ Ask an adult to bake the meringues in the bottom of a cool oven (110°C, 225°F or gas mark ¼) for 2–3 hours.

As the sugar cooks, it hardens, trapping the air inside the meringues – until you bite them!

124

EXPERIMENT Removable air

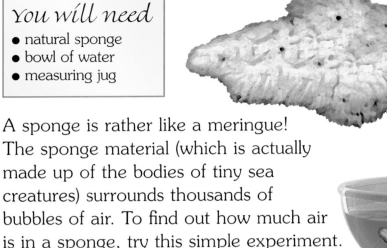

A sponge is rather like a meringue! The sponge material (which is actually made up of the bodies of tiny sea creatures) surrounds thousands of bubbles of air. To find out how much air is in a sponge, try this simple experiment.

1 First squeeze the sponge in your hands as hard as you can to push out the air and make it as small as possible.

2 Push the squeezed sponge underwater and let it go. All the spaces will fill with water.

3 Lift the sponge up quickly and squeeze it as hard as you can over a measuring jug. The volume of water in the jug is about the same as the volume of air in the sponge.

EXPERIMENT Mini-sub

1 Ask an adult to make a hole in the lid and the bottom of a plastic bottle. Push plastic tubing through the hole in the lid.

2 Put your finger over the hole in the bottom and fill the bottle to the top with water. Screw on the lid.

3 Let the bottle sink to the bottom of a large tank or bowl of water. Take your finger away from the hole and start blowing into the tube. What happens?

Glossary

Aerofoil A special shape, with a curved top surface, used in aeroplane wings to give "lift".

Aileron A flap on the wing of an aeroplane that can be moved up and down to help the aircraft to manoeuvre.

Air pressure The force with which the air presses down on things. We experience air pressure all the time. Changes in air pressure, caused by the warming of air by the Sun, makes air move and is the reason for most of the weather on Earth.

Anemometer An instrument used for measuring wind speed.

Barometer An instrument used for measuring air pressure. Readings from a barometer can give useful information about weather changes likely in the next day or so.

Condensation The changing of a gas into a liquid, usually as a result of cooling.

Humidity A measure of how much moisture (water) there is in the air.

Hygrometer A instrument used for measuring humidity (the amount of moisture in the air).

Oxygen A gas that makes up about 20% of the air we breathe. In chemical formulae, it is shown by the letter O.

Sponge A sea animal with an internal skeleton. When the animal dies, its soft, pliable skeleton forms a spongy mass, familiar in bathrooms!

Suction Causing the movement of a gas or liquid by changing the force of air pressure on it. For example, sucking reduces air pressure in a straw. Air pressure on the surface of a drink is then stronger and forces the drink up the straw.

Thermometer An instrument used for measuring temperature.

Tornado A powerful, spiralling wind created when warm, moist air meets colder, drier air moving in another direction.

Vapour A gas created by the evaporation of a liquid, such as water.

Index